2nd Edition

AP® STATISTICS
CRASH COURSE®

Michael D'Alessio, M.S.

Watchung Hills Regional High School
Warren, New Jersey

Research & Education Association

www.rea.com

Research & Education Association
1325 Franklin Ave., Suite 250
Garden City, NY 11530
Email: info@rea.com

AP® STATISTICS CRASH COURSE,® 2nd Edition

Published 2023
Copyright © 2020 by Research & Education Association.
Prior edition copyright © 2011 by Research & Education Association. All rights reserved. No part of this book may be reproduced in any form without permission of the publisher.

Printed in the United States of America

Library of Congress Control Number 2019946641

ISBN-13: 978-0-7386-1258-4
ISBN-10: 0-7386-1258-8

Cover photo © iStockphoto.com/asiseeit

AP® STATISTICS CRASH COURSE
TABLE OF CONTENTS

PART I Introduction

PART II Exploring Data

PART III Survey Methods

ABOUT OUR BOOK

REA's *AP® Statistics Crash Course®* is designed for the last-minute studier or any student who wants a quick refresher on the AP® course. The *Crash Course* is based on the the latest changes to the AP® Statistics course and exam and focuses only on the topics tested, so you can make the most of your study time.

Written by a veteran AP® Statistics test expert, our *Crash Course* gives you a concise review of the major concepts and important topics tested on the AP® Statistics exam.

- **Part I** gives you **Keys for Success**, so you can tackle the exam with confidence.

- **Part II** covers interpreting graphical displays of data, comparing distributions of univariate data and exploring bivariate data, among other topics.

- **Part III** reviews survey methods and explains data collection and planning and how to conduct surveys and experiments.

- **Parts IV–VI** cover essential information about probability and expected values, distributions of data, and statistical inference.

- **Part VII** covers specific **Test-Taking Strategies** to help you conquer the multiple-choice and free-response questions, and includes a chapter of AP®-style multiple-choice and free-response practice questions to prepare for test day.

- Also included are the **Formulas and Tables** you'll use on the exam, as well as a **Glossary** of must-know terms.

ABOUT OUR ONLINE PRACTICE EXAM

How ready are you for the AP® Statistics exam? Find out by taking **REA's online practice exam** available at *www.rea.com/studycenter*. This test features automatic scoring, detailed explanations of all answers, and diagnostic score reporting that will help you identify your strengths and weaknesses so you'll be ready on exam day.

Whether you use this book throughout the school year or as a refresher in the final weeks before the exam, REA's *Crash Course* will show you how to study efficiently and strategically, so you can boost your score.

Good luck on your AP® Statistics exam!

ABOUT OUR AUTHOR

Michael D'Alessio earned his B.S. in Biology from Seton Hall University and his M.S. in Biomedical Sciences from the University of Medicine and Dentistry of New Jersey. In 2004, he earned his Executive Master of Arts in Education Leadership from Seton Hall University.

Mr. D'Alessio has had an extensive career teaching all levels of mathematics and science, including AP® Statistics, as well as chemistry, biology, physics, algebra, calculus, and geometry. In 2003, Mr. D'Alessio received the Governor's Teacher of the Year recognition for his classroom leadership at Watchung Hills Regional High School in Warren, New Jersey.

Currently, Mr. D'Alessio serves as the Director of Science, Instructional Technology, and Assessment at Watchung Hills Regional High School.

ABOUT REA

Founded in 1959, Research & Education Association (REA) is dedicated to publishing the finest and most effective educational materials— including study guides and test preps—for students of all ages.

Today, REA's wide-ranging catalog is a leading resource for students, teachers, and other professionals. Visit *www.rea.com* to see a complete listing of all our titles.

ACKNOWLEDGMENTS

We would like to thank Larry B. Kling, Editorial Director, for his overall guidance; Pam Weston, Publisher, for setting the quality standards for production integrity and managing the publication to completion; John Cording, Technology Director, for coordinating the design and development of the REA Study Center; and Wayne Barr, Test Prep Project Manager, for shepherding this project through development and production.

We also extend our special thanks to Stu Schwartz for ensuring alignment with the current AP® Statistics exam; Mel Friedman and Stephen Hearne for technically reviewing the manuscript; Michael Porinchak, AP® Statistics teacher, Twinsburg City School District, Twinsburg, Ohio, for writing practice multiple-choice questions; Diane Goldschmidt for editorial contributions; Jennifer Calhoun for file prep; and Kathy Caratozzolo of Caragraphics, for typesetting this edition.

PART I:

INTRODUCTION

Keys for Success
on the AP® Statistics Exam

Congratulations on your decision to take AP® Statistics. Any AP® course is challenging. AP® courses represent college-level classes that you take in high school, so they are supposed to be challenging. The good news is that with your hard work and this *Crash Course* study guide, you will be guided toward success on the exam.

I. Overview

The AP® Statistics course and exam is based on the content of a non-calculus-based introductory college-level statistics course. The topics reflect major themes that are presented in college-level textbooks. To succeed on the exam, you need to master the basic concepts of statistics and apply these concepts to various situations in a traditional test format.

In 2019, the College Board organized its new course framework around nine units that provide one possible sequence for teaching the course:

Unit 1: Exploring One-Variable Data

Unit 2: Exploring Two-Variable Data

Unit 3: Collecting Data

Unit 4: Probability, Random Variables, and Probability Distributions

Unit 5: Sampling Distributions

Unit 6: Inference for Categorical Data: Proportions

Unit 7: Inference for Quantitative Data: Means

Unit 8: Inference for Categorical Data: Chi-Square

Unit 9: Inference for Quantitative Data: Slopes

In many ways, this new course framework is geared toward your teachers to guide them in planning your AP® course. But this book is written particularly for you to use as a resource throughout the school year and as a refresher in the run-up to the exam.

In the chapters that follow, you will find content-specific help for all topics covered in the College Board's current course framework, as well as tips for success and general insight into what you need to know for today's AP® Statistics exam.

To complement your AP® coursework, reviewing the corresponding *Crash Course* chapters is a great way to reinforce what you've learned in class. Here's a quick view of how this book covers the exam's new unit structure of the AP® course framework.

Where You'll Find the AP® Statistics Units in This Book

College Board AP® Course Framework	Crash Course Chapter
Unit 1: Exploring One-Variable Data	2, 3, 4, 13
Unit 2: Exploring Two-Variable Data	2, 5, 6, 7
Unit 3: Collecting Data	8, 9, 10
Unit 4: Probability, Random Variables, and Probability Distributions	7, 11, 12
Unit 5: Sampling Distributions	14
Unit 6: Inference for Categorical Data: Proportions	13, 15, 16
Unit 7: Inference for Quantitative Data: Means	15, 16
Unit 8: Inference for Categorical Data: Chi-Square	16
Unit 9: Inference for Quantitative Data: Slopes	15, 16

II. **Format and Content of the Exam**

The AP® Statistics exam is 3 hours long and consists of both a multiple-choice and free-response section.

Section	Question Type	Number of Questions	Exam Weighting	Timing
I	Multiple-choice questions	40	50%	90 minutes
II	Free-response questions	6		
	Part A: Questions 1–5		37.5%	65 minutes
	Part B: Question 6: Investigative task		12.5%	25 minutes

Section I—Multiple-choice—90 Minutes—40 questions—50% of exam grade

- Four AP® Statistics skill categories are assessed in the multiple-choice section. Though problems can address multiple skill categories, each question focuses on one of these major categories:

  Skill 1: Selecting Statistical Methods

  Skill 2: Data Analysis

  Skill 3: Using Probability and Simulation

  Skill 4: Statistical Argumentation

- Multiple-choice scores are based on the number of questions that are answered correctly.

- No points are awarded for unanswered questions.

- Because no points are deducted for incorrect answers, it is to your advantage to answer all multiple-choice questions. Try to eliminate as many answer choices as you can and then select the best answer from the remaining choices.

- Each multiple-choice question has five possible answer choices labeled (A) through (E). There is only one correct answer.

Section II—Free-Response—90 Minutes—6 questions—50% of exam grade

- Part A contains five free-response questions designed to be completed in 65 minutes and represents 37.5% of your exam grade.

  These five free-response questions assess across five areas as follows:

  1. Collecting Data—Skill 1: Selecting Statistical Methods, one multipart question

  2. Exploring Data—Skill 2: Data Analysis, one multipart question

  3. Probability and Sampling Distributions—Skill 3: Using Probability and Simulation, one multipart question

  4. Inference—Skills 1, 3, and 4: one question

  5. Multiple skill categories: one question

- Part B contains one free-response question called an Investigative Task. This investigative task assesses multiple skills, focusing on the application of skills and content in new contexts or in non-routine ways. It is designed to be completed in 25 minutes and represents 12.5% of your exam grade.

III. Scoring the Exam

The multiple-choice and free-response sections of the exam are scored independently. The multiple-choice section is scored by machine. Each of the 40 questions is worth one point and your score is based on the number of questions answered correctly. The free-response questions are scored by hand during the annual AP® Exam Reading, which takes place in June. AP® Statistics teachers and college instructors apply a scoring guide to the six free-response items and award points based on those guidelines.

The College Board uses a formula (which changes slightly from year to year) to rank your combined multiple-choice and free-response scores as a composite score. The composite score is then converted to a score on the AP® 5-point scale. This scale determines how students are to receive college credit or placement as follows:

> 5 = Extremely Well Qualified
>
> 4 = Well Qualified
>
> 3 = Qualified
>
> 2 = Possibly Qualified
>
> 1 = No Recommendation

Some colleges and universities accept scores of 3, 4, or 5 for college credit or placement, while others accept only a 4 or 5. Some colleges do not award credit for an AP® exam, so research the policies of the colleges you are interested in attending. Also, be aware that colleges and universities can change their AP® acceptance policies at any time. Stay up-to-date by checking the latest AP® policies on their websites.

Even if a college does not award credit for an AP® test, taking an AP® exam may strengthen your college application because you rose to the "AP® challenge."

IV. Calculator Usage

Graphing calculators with statistical capabilities can be used during the exam and students are expected to bring their own. Although you can do well on the exam without a calculator, not having one would be a disadvantage.

Knowing and understanding the power of a graphing calculator is essential for success on the AP® Statistics exam. If not provided by your school, you should invest in a TI-83, TI-83+, TI-89, or TI-Nspire calculator. This type of calculator will assist you throughout your high school and college career as well as on the AP® Statistics exam. Unlike other mathematics or science AP® tests, you may use the graphing calculator throughout the entire test. The following are some functions you should be familiar with on the graphing calculator (both univariate and bivariate data):

- Summary Statistics (mean, median, mode, standard deviation, quartiles)
- Histogram Plot
- Boxplots
- Scatterplots
- Least Square Regression Lines
- Residuals
- Probability
- Simulation
- Confidence Intervals
- Tests of Significance

Check the AP® Statistics course page (*apstudents.collegeboard. org*) for the most up-to-date list of eligible calculators and policies.

V. Formulas and Tables

A list of formulas and tables is furnished to AP® Statistics examinees at the test site. Familiarize yourself with these formulas and tables, which can be found in Appendix A of this book.

VI. Supplementing Your *Crash Course*

This *Crash Course* contains what you need to know to score well on the AP® Statistics exam. You should, however, supplement it with materials provided by the College Board such as the AP® Statistics Course and Exam Description. The College Board's AP® Central website contains a wealth of materials, including free-response questions with exemplars and rubrics.

VII. Exam Day

1. Arrive at the exam site at least 20 minutes before the scheduled start time for the exam.

2. Bring two fresh No. 2 pencils with clean erasers and two working blue- or black-ink pens.

3. Bring a snack and a bottle of water for the short break between the two sections of the exam.

4. Bring your graphing calculator.

5. The exam proctor will read a lot of instructions—be patient. Plan to spend 3 to 4 hours at the exam site.

6. Answer *every* multiple-choice question, even if you have to guess. Remember, there is no deduction for an incorrect answer, but no points can be earned for unanswered questions. If you're stuck, give it your best guess.

7. When you're done, relax. Sit back and wait for your qualifying score to arrive!

PART II:

EXPLORING DATA

Exploring Data:
Constructing and Interpreting Graphical Displays of Data

I. Key Terms

A. Sample Survey—a statistical method of sampling a large population of individuals in order to draw conclusions from questions. Responses (data) to a survey are recorded and then analyzed through various statistical methods.

Example

Do more people enjoy watching Major League Baseball during the day or evening hours?

B. Experiment—individuals are subjected to some type of treatment. Results (data) are recorded scientifically and are analyzed statistically.

Example

Does the addition of Vitamin C to a person's diet reduce the chances of catching the common cold?

C. Individuals—objects in the data set.

Example

People, animals, a sports team, countries.

D. Variable—a characteristic of objects in a data set.

Example

The SAT scores of 11th graders at Thomas Jefferson Memorial High School.

E. Categorical (also called qualitative) Variable—places an individual into one of several groups/categories.

Example

Animals that hibernate in the winter.

F. Quantitative Variable—a variable that takes on a numerical variable.

Example

The grade point averages of students during their first semester at Eastern University.

NBA Team	Region of United States	Average Yearly Payroll (2000–2010)	Number of Championships (2000–2010)
New York Knicks	East	$52,000,000	0
Los Angeles Lakers	West	$61,000,000	5
San Antonio Spurs	Southwest	$48,500,000	3

➤ In the example above, the individuals are the different NBA teams.

➤ In the example above, the Region of the United States is a categorical variable.

➤ In the example above, the Average Yearly Payroll is a quantitative variable.

➤ In the example above, the Number of Championships is a quantitative variable.

G. Distribution—indicates the value of the variable and its frequency of taking place.

Example

Sampling 1,000 high school students about their favorite subject (Mathematics, English, Social Studies, etc.).

Favorite Subject	Frequency
English	230
Math	160
Social Studies	180
Science	115
Art	140
Music	175

II. Describing Data

A. Bar Graph—used for a qualitative (non-number based) independent variable (*x*-axis).

Percent of Boston Red Sox Fans in 4 New England States in 1973

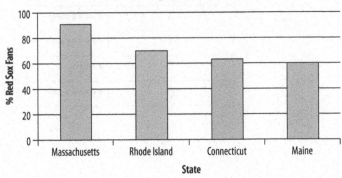

B. Side-by-Side Bar Graph—allows for a more in-depth analysis using the qualitative independent variable.

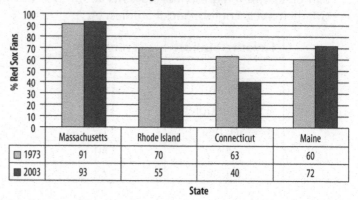

Percent of Boston Red Sox Fans in 4 New England States in 1973 & 2003

	Massachusetts	Rhode Island	Connecticut	Maine
☐ 1973	91	70	63	60
■ 2003	93	55	40	72

State

C. Dotplot—used to depict quantitative data in a quick and easy manner.

Single Die Roll	Frequency
1	●●●●●●●●
2	●●●●●●●●●●
3	●●●●●●●●●
4	●●●●●●●●●●●●
5	●●●●●
6	●●●●●●●●●

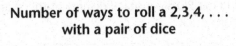

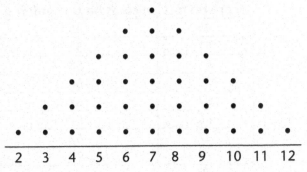

**Number of ways to roll a 2,3,4, . . .
with a pair of dice**

D. Pie Charts—used to relate each category to the whole (not as flexible or as common as a bar graph). Based on 360 degrees of a circle.

Percent of Car Color for Toyota 4x4 Sales, 2009

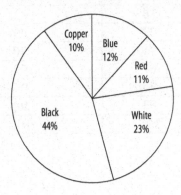

E. Stem and Leaf Plots—a quick display of the shape of the distribution.

Score on test	Scores on test ordered
45	45
51	51
67	58
77	66
87	67
66	67
67	67
78	76
79	77
79	78
81	79
90	79
92	81
89	82
58	85
76	87
82	89
85	90
67	92
96	96
98	98
100	100

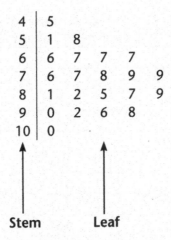

```
 4 | 5
 5 | 1  8
 6 | 6  7  7  7
 7 | 6  7  8  9  9
 8 | 1  2  5  7  9
 9 | 0  2  6  8
10 | 0
```

Stem   **Leaf**

The stem and leaf plot above indicates a somewhat symmetric shape. Outline the leaf and identify the shape when constructing such graphs.

Steps in Constructing a Stem and Leaf Plot

1. Order the data from smallest to largest.

2. The stem consists of all the left-most digits EXCEPT the final digit.

3. The leaf contains the final digit.

4. The stem is vertical from smallest to largest value descending.

5. Draw a vertical line between the stem and leaf.

6. Write each leaf in a row to the right of the stem.

Stem and Leaf Range Plot

County	Median Household Income In Thousands	Ordered
Somerset	$77	$39
Morris	$77	$40
Hunterdon	$80	$42
Bergen	$65	$44
Monmouth	$64	$45
Mercer	$64	$46
Sussex	$65	$46
Union	$55	$48
Middlesex	$61	$49
Burlington	$59	$54
Warren	$56	$55
Essex	$45	$56
Cape May	$42	$59
Ocean	$46	$61
Gloucester	$54	$64
Camden	$48	$64
Passaic	$49	$65
Hudson	$40	$65
Atlantic	$44	$77
Salem	$46	$77
Cumberland	$39	$80

```
3 | 9
4 | 0  2  4  5  6  6  8  9
5 | 4  5  6  9
6 | 1  4  4  5  5
7 | 7  7
8 | 0
```

Back-to-Back Stem and Leaf Plot Scores of Students on an AP® Statistics Test in Two Different Classes

Mrs. Jones				Stem		Mrs. Smith	
				2			
			0	3			
			2	4			
				5	2	2	5
7	6	5	1	6	1	2	3
				7			
		2	2	8	5	6	8
			1	9	4		
			0	10	1		

Test Tip

Construction and interpretation of a stem and leaf plot is highly likely to be on the AP® Statistics exam. Practice using old free-response questions from the College Board website. Be sure to download the scoring guidelines to see how you are doing.

F. Frequency Bar Graph—displays the count of percent of observations that fall into a certain class.

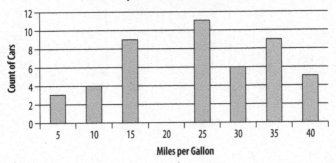

Bar Graph of MPG for Random Cars

G. Cumulative Frequency Plot—displays cumulative information graphically. The number, percentage, or proportion of observations in a distribution that are less than or equal to particular values.

Graph A

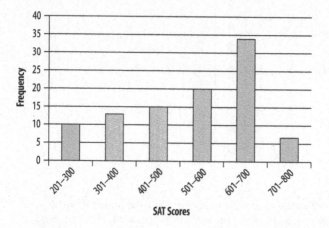

Graph A depicts a normal bar graph where 20 students earned a score of 501–600 on the SAT.

Graph B

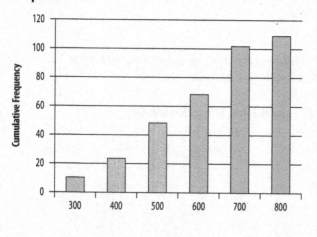

Graph B depicts a cumulative frequency plot that indicates up to and including each test score. For example, roughly 70 students scored 600 or less on the SAT in this sample.

H. Histograms

1. A popular pictorial method that is similar to a bar graph.

2. A histogram is a graph that contains vertical connected rectangular bars in the first quadrant of the *xy*-coordinate plane.

3. The given data is grouped into classes and each bar shows the frequency of a particular class.

Example

The following frequency distribution shows the record low temperatures of last year for twenty-five cities.

Class	Frequency
3°–9°	3
10°–16°	5
17°–23°	9
24°–30°	6
31°–37°	2

For each class, there is a lower limit and an upper limit. In the first class, the lower limit is 3° and the upper limit is 9°. To construct the histogram, identify the lower and upper boundary of each class. Each temperature represents a continuous (as opposed to discrete) number. For example, 3° could represent any number between 2.5° and 3.5°. Similarly 9° could represent any number between 8.5° and 9.5°. To create the first bar of the histogram, use the lower boundary of the lower limit and upper boundary of the upper limit of the first class. This means that the first bar will begin with the number 2.5° and end with the number 9.5°.

Note that the lower and upper limits of the second class are 10° and 16°, respectively. The lower boundary for 10° is 9.5° and the upper boundary for 16° is 16.5°. Thus, the second bar will begin with 9.5° and end with 16.5°. Notice that the upper boundary of the first class is identical to the lower boundary of the second class.

By continuing in this manner, here are the upper and lower boundaries for the third, fourth and fifth classes:

Third class: lower boundary = 16.5° and upper boundary = 23.5°

Fourth class: lower boundary = 23.5° and upper boundary = 30.5°

Fifth class: lower boundary = 30.5° and upper boundary = 37.5°

The width of each rectangular bar is the difference between its upper and lower boundary. Note that this common width for all these bars is 7°. The complete histogram appears below.

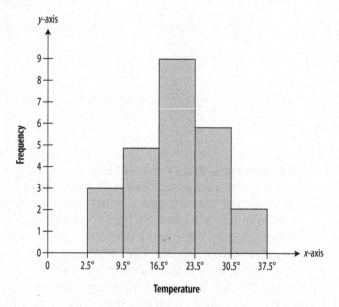

III. Categorizing a Distribution

A. Pattern—shape, center, and spread

1. Symmetric Shape—the values to the left (smaller) and to the right (larger) are mirror images of each other.

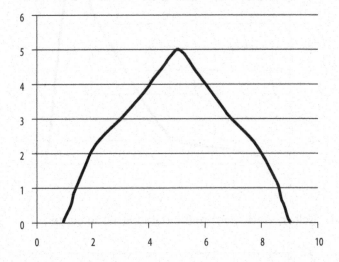

2. Skewed to the right—the data spreads far and thin toward the higher values.

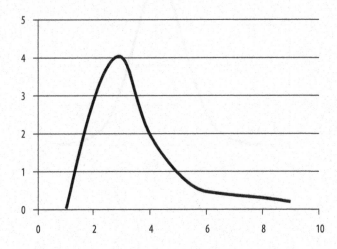

3. Skewed to the left—the data spreads far and thin toward the lower values.

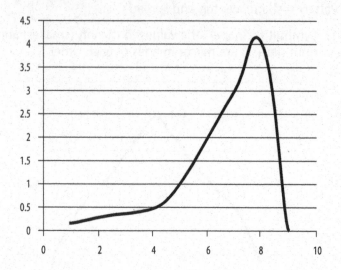

4. Bell-shaped—symmetric data with one center.

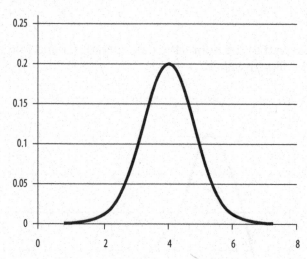

5. Uniform—a histogram with a horizontal line.

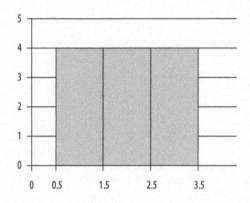

6. Modes—number of peaks the distribution contains.

    i.   Unimodal—a distribution with only one peak.

    ii. Bimodal—a distribution with two peaks.

B. Center—portion of the distribution that separates the distribution roughly in half.

C. Spread—the values of the distribution from smallest to largest.

D. Gaps—when the distribution has no value at a certain point.

E. Outlier—individual that falls outside of the overall pattern of data

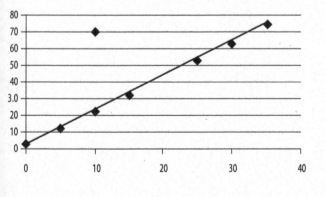

The data point (10,70) represents an outlier

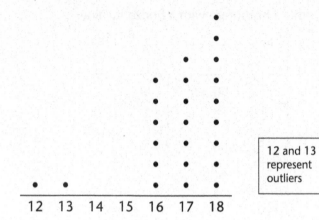

12 and 13 represent outliers

Don't be the OUTLIER test-taker who does not know anything about OUTLIERS. Free-response questions may have you state the outlier and offer reasons for its presence in a data set. Reasons for outliers include: human error, fraudulent error, device/instrument error (your ruler is broken), or error in communication (the data was not transformed correctly).

Exploring Data:
Describing Distributions with Numbers

I. Measure of Center: Mean and Median

A. Mean (or arithmetic mean) is defined as the average value.

1. Formula for mean $\bar{x} = \dfrac{x_1 + x_2 + \ldots + x_n}{n} = \dfrac{\sum x_i}{n}$

   $\bar{x}$ = mean

   $\sum$ = sum of

   x_i = total observations

2. The mean is extremely sensitive to outlier data. Outliers can change the location of the mean greatly. Therefore, the mean is not a resistant measure of center.

Example

A) The mean of the following numbers can be determined using the formula above.

   4, 6, 2, 5, 7

   $$\bar{x} = \frac{4+6+2+5+7}{5} = 4.8$$

B) If an outlier such as 30 is added to the list above, then a new mean is calculated.

   $$\bar{x} = \frac{4+6+2+5+7+30}{6} = 9$$

B. The median (*M*) is the midpoint of any distribution set or the number in which half the observations are smaller than or equal to *M* and half the observations are larger than *M*.

1. Rules to finding the median.

   i. Arrange all observations from smallest to largest.

   ii. When the number of observations is odd the median *M* is found by adding 1 to the number of observations and dividing by 2 or $\frac{n+1}{2}$ from the bottom of the list.

   iii. When the number of observations is even, the median *M* is the average of the two center observations in the ordered list. The median is the average of the $\frac{n}{2}$ and $\frac{n}{2}+1$ observations from the bottom of the list.

   Example

Year	Points Per Game Average for Player X (Rounded to the nearest whole number)
1986	23
1987	37
1988	35
1989	33
1990	34
1991	31
1992	35
1993	33
1994	27
1995	31
1996	26
1997	29
1998	23
1999	20
2000	10

Example

Find the median point per game for Player X.

To find the median, arrange the numbers from least to greatest.

10, 20, 23, 23, 26, 27, 29, 31, 31, 33, 33, 34, 35, 35, 37

Since there are 15 observations, the median would be the middle value from the left. Using $\frac{n+1}{2}$ we find that $\frac{15+1}{2} = 8$ or the 8th term (*not the value 8*). The median in this case would be 31.

Example

Determine the median points per game for the player excluding the 2000 season (in that year the player only played 5 games due to injury).

Excluding 2000, arrange from least to greatest.

20, 23, 23, 26, 27, 29, 31, 31, 33, 33, 34, 35, 35, 37

Since there are 14 observations, the median is the average of the two middle terms. The location of the middle terms are as follows:

$$\frac{n}{2} = \frac{14}{2} = 7$$

$$\frac{n}{2} + 1 = 7 + 1 = 8$$

Take the average of the 7th and 8th terms, which is $\frac{31+31}{2} = 31$.

From the example above, you can observe the median is a *more* reliable measure of center than the mean. Removing the year, 2000, did not change the median at all.

C. Some important facts about mean and median.

1. If the distribution is symmetric, the mean and the median will be close or exactly the same (if the distribution is exactly symmetric).

2. If the distribution is skewed, the mean will be in the tail section, farther out than the median.

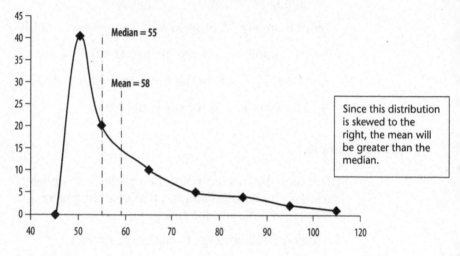

Since this distribution is skewed to the right, the mean will be greater than the median.

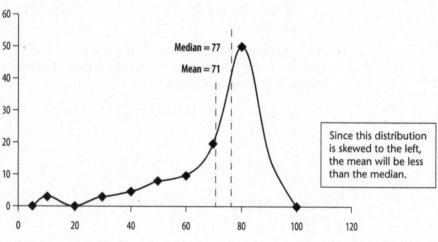

Since this distribution is skewed to the left, the mean will be less than the median.

II. Measuring Spread: Range, Interquartile Range, Standard Deviation

A. Range—the difference between the largest and smallest observations.

B. Percentiles—the nth percentile of any distribution is the value such that n% of the observations fall at or below it. For example, if a student scores in the 90th percentile on the SAT mathematics section, this indicates that the student scored in the top 10% and better than 90% of the students who took the test.

1. In statistics the most common percentiles are called quartiles.

   i. The first quartile, Q_1, is the 25th percentile.

   ii. The second quartile is equal to the median, the 50th percentile.

   iii. The third quartile, Q_3, is the 75th percentile.

2. Calculating Quartiles

   i. Arrange all observations from least to greatest.

   ii. To calculate Q_1 find the median of the distribution that is to the *left* of the overall median.

   iii. To calculate Q_3 find the median of the distribution that is to the *right* of the overall median.

   iv. Interquartile Range—the difference between the first and third quartiles: $IQR = Q_3 - Q_1$

   Example (even number of terms):

   Find Q_1, the median, and Q_3 for the data below showing the number of home runs hit by a baseball player over a 10-year period (data arranged in order).

   9, 10, 11, 13, 14, 15, 15, 16, 17, 18

Median is equal to the average of the 5th and 6th terms:
$\frac{14+15}{2} = 14.5$.

14.5

9 10 11 13 14 | 15 15 16 17 18

Q_1 = The median of the first five terms or 11

Q_3 = The median of the last five terms or 16

Example (odd number of terms):

Find Q_1, the median, and Q_3 for the data below about the number of home runs hit by a baseball player over a 9-year period (data arranged in order).

30, 31, 33, 34, 34, 35, 37, 41, 44

Median is equal to the 5th term: 34.

30 31 33 34 34 35 37 41 44

Q_1 = The median of the first four terms or $\frac{31+33}{2} = 32$

Q_3 = The median of the last five terms or $\frac{37+41}{2} = 39$

III. Constructing a Boxplot

A. Uses the five-number summary of data (Minimum, Q_1, M, Q_3, Maximum).

1. A central box spans Q_1 and Q_3.

2. A line in the box marks the median.

3. Lines extend from the box out to the smallest and largest observations.

B. Used for comparisons of more than 1 distribution.

C. The quartiles show the spread of half the data.

Example:

Age of 30 Random People in a Shopping Mall				
23	23	23	25	25
25	26	27	27	28
33	34	36	40	41
42	42	43	45	46
47	48	50	55	56
70	71	72	73	74

Minimum	Q_1	Median	Q_3	Maximum
23	27	41.5	50	74

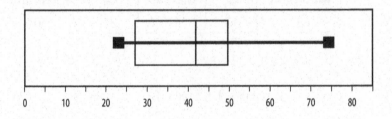

D. *1.5 × IQR Rule*—a rule used for detecting suspected outliers. If an observation falls more than *1.5 × IQR* above Q_3 or below Q_1, then that observation is a suspected outlier.

Example

Brady Anderson was a center fielder for the Baltimore Orioles from 1988 to 2001. His home run total for Baltimore is summarized below for the years 1992 to 2001. Which years are possible outliers based on the *1.5 × IQR* rule?

1992	1993	1994	1995	1996	1997	1998	1999	2000	2001
21	13	12	16	50	18	18	24	19	8

Minimum	Q_1	Median	Q_3	Maximum
8	13	18	21	50

$$\bar{x} = 20, \; IQR = 8$$

$$1.5 \times IQR = 1.5 \times 8 = 12$$

$$Q_1 - 1.5 \times IQR = 13 - 12 = 1$$
$$Q_3 + 1.5 \times IQR = 21 + 12 = 33$$

From the data, it is confirmed that 1996 or 50 home runs is an outlier of the data.

IV. Measuring Spread: Standard Deviation

A. Variance or s^2 is the average of the square of the deviations of the data from their mean.

Formula for Variance: $s^2 = \dfrac{1}{n-1}\sum(x_i - \bar{x})^2$

B. Standard Deviation measures spread by looking at how far the observations are from the mean. The standard deviation is the square root of the variance (s^2).

1. Standard deviation measures the spread about the mean, and is used when the mean is chosen as the measure of center.

2. When the standard deviation equals zero, there is no spread.

3. The standard deviation is greatly affected by outliers.

Formula for Standard Deviation: $s = \sqrt{\dfrac{1}{n-1}\sum(x_i - \bar{x})^2}$

n = number of observations

Example

Find the standard deviation of the data below.

Age of the 5 starting pitchers of the New York Mets:

23, 31, 24, 28, 27

$\bar{x} = 26.6$

$s = \sqrt{\dfrac{1}{4}(26.6-23)^2 + (26.6-31)^2 + (26.6-24)^2 + (26.6-28)^2 + (26.6-27)^2}$

$s = 3.21$

A standard deviation of 3.21 indicates that these observations have an average spread of 3.21 units from the mean.

C. *Z-score*—a numerical value that indicates how many standard deviations (and in which direction) away from the mean the original observation falls. Calculating a *z*-score is also called "standardizing."

Formula for *z*-score: $z = \dfrac{x - mean}{s}$

s = standard deviation

x = original observation

Example

Below is the number of offensive touchdowns scored by the 11 teams that comprise the Big Ten Football Conference (2009 data).

Team	Offensive Touchdowns
Wisconsin	52
Michigan State	47
Michigan	46
Ohio State	44
Penn State	47
Purdue	43
Northwestern	40
Illinois	38
Indiana	34
Iowa	35
Minnesota	33

Using a graphing calculator, the following Summary Statistics were found:

Min = 33

$Q_1 = 35$

Med = 43

$Q_3 = 47$

Max = 52

$\bar{x} = 41.7$

$s = 6.19$

What was the standardized touchdown score or *z*-score for Illinois and Ohio State?

Illinois $\quad z = \dfrac{x - mean}{s} = \dfrac{38 - 41.7}{6.19} = -0.60$

Ohio State $\quad z = \dfrac{x - mean}{s} = \dfrac{44 - 41.7}{6.19} = 0.37$

In other words, Illinois is 0.60 standard deviations below the mean and Ohio State is 0.37 standard deviations above the mean.

D. The Effect of Changing Units on Univariate Data

1. Linear Transformation—a change in the units of measurement being used. For example, changing pounds to kilograms or miles to kilometers.

   Formula for Linear Transformation:

   $x_{new} = a + bx$

   x_{new} = new variable

   a = up or down shifts

   b = changing the size of the unit

   i. When multiplying each observation by "*b*," the mean, median, IQR, and standard deviation can also be multiplied by b.

   ii. Adding "*a*" to each observation changes the mean and median, but does not change spread measures such as IQR or standard deviation.

   iii. Linear transformations do not change the shape of the distribution.

Below is a summary relationship between miles and kilometers. To convert miles to kilometers simply multiply the number of miles by 1.61. The mean, median, and standard deviation change by a factor of 1.61.

When 1 extra kilometer is added, the mean and median change by 1, but the standard deviation stays the same.

	Miles	$x_{new} = 0 + (1.61)(miles)$ kM	$x_{new} = 1 + (1.61)(miles)$ Add 1 kM
	4.1	6.601	7.601
	3.4	5.474	6.474
	2.7	4.347	5.347
	4.2	6.762	7.762
	5.1	8.211	9.211
Mean	3.90	6.28	7.28
Median	4.10	6.60	7.60
Standard Deviation	0.90	1.45	1.45

Test Tip

When you analyze one-variable data (univariate), always discuss center, shape, and spread. Look for patterns in the data, and then for deviations from those patterns—outliers.

When commenting on center: Median and mean are not the same. They are both measures of center, but for a given data set, they may differ.

When commenting on shape: Symmetric is not the same as "equally" or "uniformly" distributed. Do not say that a distribution is "normal" just because it looks symmetric and unimodal. Skewness can be to the left or the right.

When commenting on spread: Variance and standard deviation are not the same measurement. Standard deviation is the square root of the variance.

Comparing Distributions of Univariate Data
(Dotplots, Back-to-Back Stemplots, Parallel Boxplots)

I. The Most Efficient Way to Study Distributions

A. Who are the individuals and what data is being analyzed?

B. What types of graphs would best describe the data?

C. What are the summary statistics?

D. Discuss center, spread, clusters, gaps, outliers, and shapes.

II. Dotplots

Example

The spending habits of two different groups of college students were recorded using dotplot graphs. Group A were college students who had jobs while going to school. Group B college students did not have jobs while going to school. The numbers indicate the total dollar value spent per week.

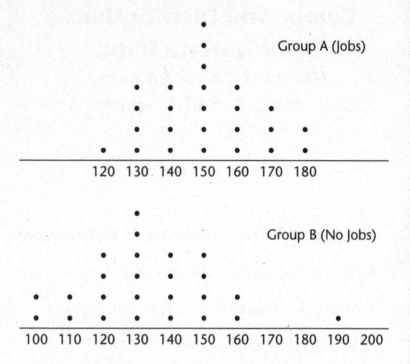

➤ Who? College students who had or did not have a job during the school year.

➤ What? The amount of money they spent per week.

➤ What types of graphs should be used? Dotplots.

➤ Summary Statistics (on calculator)

Group A

```
1-Var Stats
  x̄ = 150.4
  Σx = 3760
  Σx² = 571800
  5x = 16.19670748
  σx = 15.86946754
↓n = 25
■
```

```
1-Var Stats
↑n = 25
  minX = 120
  Q₁ = 140
  Med = 150
  Q₃ = 160
  maxX = 180
```

Group B

```
1-Var Stats
  x̄ = 134
  Σx = 3350
  Σx² = 458900
  5x = 20.41241452
  σx = 15.86946754
↓n = 25
```

```
1-Var Stats
↑n = 25
  minX = 100
  Q₁ = 120
  Med = 130
  Q₃ = 150
  maxX = 190
```

➤ Center, spread, clusters, gaps, outliers, and shapes.

- The center of group A is 150 and group B 130. Both of these correspond closely to their respective mean and median.

- The spread of the data for group B (100–190) is bigger than that of group A (120–180). The standard deviations for the two distributions are close (15.86 vs. 20).

- The shape of the group A graph is bell-shaped, while that of group B is also bell-shaped, with one outlier measure at 190.

- Using $1.5 \times IQR$
    $1.5 \times 30 = 45$
    $Q_3 + 45 = 150 + 45 = 195$

Technically speaking, 190 is not an outlier, but it is close!

Back-to-Back Stemplot

Two distinct groups of mice were treated with a placebo or an anti-cancer drug, and the number of months the mice lived was plotted back to back.

Mice w/ Placebo		Mice w/ Anti-Cancer Drug
	1 | 0	
	2 | 1	2
7 6 5 5 5 4 1 | 2	5 8 9
9 9 8 7 7 4 3 | 3	3 4 5 5 5 7
	4 | 4	7 8 8 9 9

➤ Who? Mice treated with a placebo or anti-cancer drug.

➤ What? Lifespan in months.

➤ What types of graphs should be used? Back-to-back stemplots.

➤ Summary Statistics (on calculator)

Mice with placebo

```
1-Var Stats
 x̄ = 27.6875
 Σx = 443
 Σx² = 13911
 5x = 10.47357787
 σx = 10.14099816
↓n = 16
▪
```

```
1-Var Stats
↑n = 16
 minX = 1
 Q₁ = 24.5
 Med = 26.5
 Q₃ = 37
 maxX = 39
```

Mice with anti-cancer drug

```
1-Var Stats
 x̄ = 36.75
 Σx = 588
 Σx² = 23238
 5x = 10.42113238
 σx = 10.09021804
↓n = 16
```

```
1-Var Stats
↑n = 16
 minX = 12
 Q₁ = 31
 Med = 35
 Q₃ = 47.5
 maxX = 49
```

➤ Center, spread, clusters, gaps, outliers, and shapes.

- The center of the placebo group is 26.5 and the center of the drug group is 35. Both of these correspond closely to their respective mean and median.

- The range of the data for the placebo group is 38 and for the drug group is 37. The standard deviations for the two distributions are roughly the same.

- The placebo group tends to cluster from 20–40 months, while the drug-treated group's cluster is spread to 25–50 months, indicating some efficacy for the drug versus the placebo. Standard deviations are roughly the same.

- Both sets of data indicate a very low skewness, since the mean and median are close to one another.

Test Tip

Every released AP® Statistics test has contained a back-to-back stemplot on it. The exam you take will be no different and it will ask you a series of questions relating to summary statistics as well as center, shape, and spread.

Boxplots

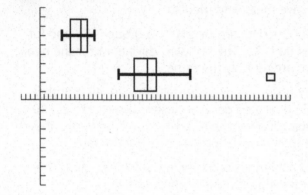

Modified boxplot that shows outliers.

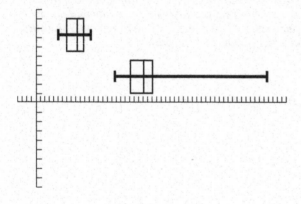

Boxplot that incorporates the outlier.

Parallel Boxplots

- The same data from the experiment with mice was plotted with parallel boxplots.

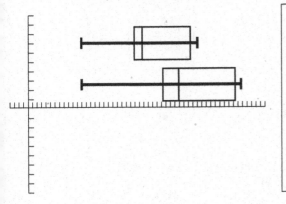

The boxplot(s) show the minimum, maximum, median, Q_1, and Q_3.

Minimum to Q_1 =
    25% of total data

Q_1 to median =
    25% of total data

Median to Q_3 =
    25% of total data

Q_3 to Maximum =
    25% of total data

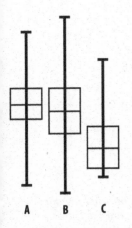

A   B   C

In the 3 parallel boxplot(s) to the left, plot A and B have roughly the same center, which are both bigger than C. Plot B has a large variable. Plot A and B are symmetric, but plot C is skewed towards the right. There are no outliers.

Plot A

```
1-Var Stats
x̄ = 9.3
Σx = 93
Σx² = 909
Sx = 2.213594362
σx = 2.1
↓n = 10
```

```
1-Var Stats
↑n = 10
 minX = 6
 Q₁ = 8
 Med = 9
 Q₃ = 10
 maxX = 14
■
```

Plot B

```
1-Var Stats
x̄ = 8.5
Σx = 85
Σx² = 809
Sx = 3.100179206
σx = 2.941088234
↓n = 10
```

```
1-Var Stats
↑n = 10
 minX = 3
 Q₁ = 7
 Med = 8.5
 Q₃ = 10
 maxX = 15
■
```

Plot C

```
1-Var Stats
x̄ = 6.388888889
Σx = 57.5
Σx² = 415.25
Sx = 2.446653043
σx = 2.30672661
↓n = 9
```

```
1-Var Stats
↑n = 9
 minX = 4
 Q₁ = 4.75
 Med = 6
 Q₃ = 7.5
 maxX = 12
```

Test Tip

Don't be surprised if a combination question with any one of the three plots in this chapter with a histogram shows up on the test. You may be asked to compare boxplots and histograms.

Exploring Bivariate Data
(Part 1)

I. Bivariate Data

Bivariate data is the study of data between two different variables. For example, SAT scores and GPA, exercise time versus heart rate, index finger length versus palm size.

A. Response variable or dependent variable measures the outcome of any study and is plotted on the *y*-axis.

B. Explanatory variable or independent variable influences the response variable and is plotted on the *x*-axis.

C. Scatterplots are graphs that are used to find relationships in bivariate data. More specifically, scatterplots are used to find the relationship between two quantitative variables measured on the same individuals.

1. Direction of a scatterplot—the overall patterns move up left to right (positive association) or down left to right (negative association).

2. Form of a scatterplot—linear relationships, but can also be curved or clustered.

3. Strength of a scatterplot—how closely the points in the scatterplot are to a straight line.

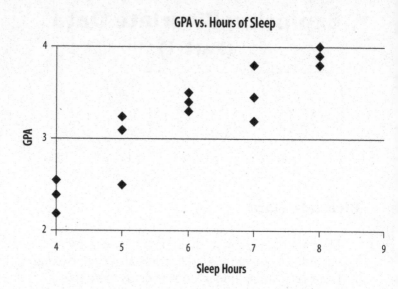

> The graph of GPA vs. Hours of Sleep has a positive correlation. It is relatively linear, indicating high positive correlation.

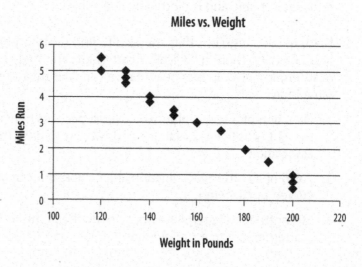

> The graph of Miles vs. Weight has a negative correlation. It is relatively linear, indicating high negative correlation.

Pay vs. Experience

➤ The graph of Pay vs. Experience has a strong positive correlation with one outlier.

II. Correlation Coefficient

The correlation coefficient (r) measures the directions and strength of the linear relationship between two quantitative variables. Correlation is not a resistant measure of center and outliers can influence its calculation.

A. The formula for correlation is below:

Use your calculator or software package to find r.

$$r = \frac{1}{n-1}\sum\left(\frac{x_i - \bar{x}}{s_x}\right)\left(\frac{y_i - \bar{y}}{s_y}\right)$$

$x_i = x$ value for ith individual

$y_i = y$ value for ith individual

$\bar{x} =$ mean for x values

$\bar{y} =$ mean for y values

s_x = st. deviation for *x* values

s_y = st. deviation for *y* values

1st individuals would be x_1 and y_1

2nd individuals would be x_2 and y_2 and so on

➤ Patterns close to 1 or –1 are closer to a straight line and thus correlate the <u>*strength of the linear relationship*</u> between quantitative values only.

B. Examples of scatterplots with correlation values.

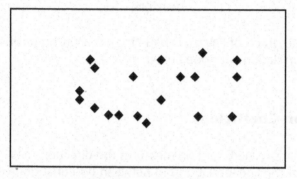

➤ No correlation, *r* = 0

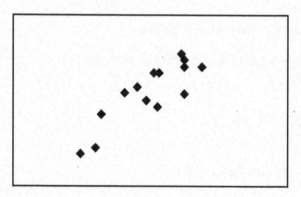

➤ Positive Correlation, *r* = +0.86

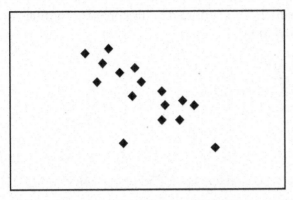

➤ Negative Correlation, $r = -0.71$

The line of best fit that rises quickly from left to right is called a positive correlation. The line of best fit that falls quickly from left to right is called a negative correlation. Strong positive and negative correlations have data points very close to the line of best fit. Weak positive and negative correlations have data points that are not clustered near or on the line of best fit. Data points that are not close to the line of best fit are called outliers.

C. Using your calculator for correlation calculation.

Enter the following data in L_1 and L_2 of your calculator and use the command 2-Var Stats under the STAT button.

0	1	2	3	4
0	.5	1.8	2.5	2.9

```
2-Var Stats
 x̄ = 2
 Σx = 10
 Σx² = 30
 Sx = 1.58113883
 σx = 1.414213562
↓n = 5
```

```
2-Var Stats
↑σy = 1.121784293
 Σxy = 23.2
 minX = 0
 maxX = 4
 minY = 0
 maxY = 2.9
```

➤ Define L_3 and L_4 (based on the previous formula for correlation)

```
((L₁-x̄)/(5x)→L₃:(
(L₂-ȳ)/5y)→L4
{-1.227881227-...
```

Special characters are found under VARS>5: STATISTICS

➤ Define L_5 as $L_3 * L_4$

L3	L4	L5 5
-1.265	-1.228	1.5532
-.6325	-.8292	.52444
0	.2073	0
.63246	.76543	.4841
1.2649	1.0844	1.3716
------	------	------
L5 = L3*L4		

➤ Find the correlation using this string: $(1/(5-1)*\text{sum}(L_5)$

```
(1/(5-1))*sum(L5)
          .983332166
```

Sum is located under LIST>MATH>5:sum(

For the data $r = 0.98$ indicating a strong positive correlation.

III. Regression Lines

A. Regression Line—a line that explains how the value of *y* changes with respect to *x*. A regression line will predict a value of *y* based on a known value of *x*.

B. Regression lines have the equation $y = a + bx$, b = slope, a = *y*-intercept.

    1. Using the GPA vs. Hours of Sleep graph, a regression line can be found that fits the model.

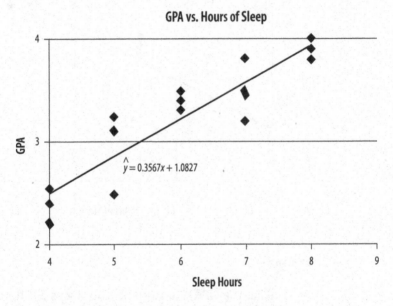

GPA vs. Hours of Sleep

➤ The regression line $\hat{y} = 0.3567x + 1.0827$ can be translated the following way: GPA = 1.0827 + 0.3567 (sleep change).

➤ *b* or the slope = 0.3567 which means that GPA increases 0.3567 with every hour of sleep.

➤ *a* or the *y*-intercept = 1.0827 which means GPA will increase 1.0857 when sleep hours do not change.

C. Predicting from a linear fit

1. Using the graph—if no regression line equation is given, use the "up and over technique." Find the value of interest on the x-axis and go up to the line and over to the y-axis.

2. Using the regression line—simply plug the value of interest on the x-axis into the equation and solve for the corresponding y-value.

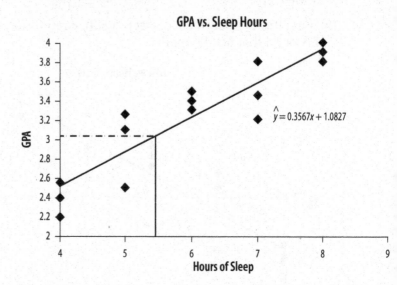

GPA vs. Sleep Hours

$\hat{y} = 0.3567x + 1.0827$

Example

➤ Predict the GPA of a student who sleeps 5.4 hours using the "up and over technique" and the regression equation.

➤ Up and over is approximately 3.00 (look at the dashed lines on the graph).

➤ Use the linear regression equation.

$$\hat{y} = 0.3567x + 1.0827$$

$$\hat{y} = 0.3567(5.4) + 1.0827 = 3.00$$

D. Extrapolation—making use of the regression line for a prediction outside the range of the graph. Often, but not always, inaccurate.

IV. The Least Squares Regression Line

A. No line drawn through any scatterplot will connect all the points of the plot.

B. Hand drawing of a regression line could be different for each person who is drawing the line (no two lines will be drawn the same).

C. A regression line is deemed adequate when the vertical distances between the points of the scatterplot and regression line are at a minimum.

D. The least squares regression line for a set of data depends on the means and standard deviations for a set of points with respect to *x* and their correlation *r*. The formula for least squares regression line is:

$$\hat{y} = a + bx$$

with slope $b = r \dfrac{s_y}{s_x}$

that passes through $(\bar{x}, \bar{y})$

Example

Data about the height of a basketball player in inches versus number of minutes played was recorded for 21 different players. Find the regression line that best fits the data.

Height	Minutes
72	47
72	38
72	42
73	42
73	42
74	40
75	38
75	38
76	35
76	35
76	33
77	32
77	30
78	30
78	35
79	29
80	28
81	29
82	26
84	22
84	29

Mean Height	76.9
Mean Minutes	34.3
Std Dev Height	3.745
Std Dev Minutes	6.333
r (correlation)	−0.912

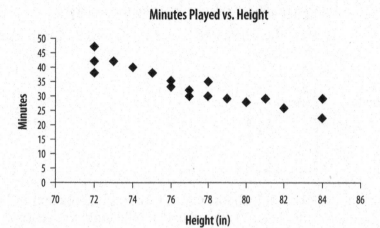

Calculation:

$$b = r\frac{s_y}{s_x} = -0.912\left(\frac{6.333}{3.745}\right) = -1.542\frac{min}{in}$$

$$\hat{y} = a + bx$$

$$34.3 = a - 1.542(76.9)$$

$$34.3 = a - 118.6$$

$$a = 152.9$$

The least squares equation is: $\hat{y} = 152.9 - 1.54x$

Using a graphing calculator, verify the results:

1) Enter the y data into L_1 and x data into L_2. Graph a scatterplot of the data.

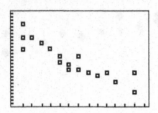

2) Press STAT, choose CALC, then 8: Lin Reg(a+bx) and then enter $L_1, L_2,$ Y1 (found under VARS button select (Y-VARS) and then ENTER.

```
LinReg
  y = ax+b
  a = -1.543788187
  b = 152.9368635
  r² = .8334696909
  r = .9129456122
```

3) Enter the equations into the Y=screen and press GRAPH. Use ZoomStat.

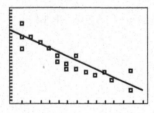

V. Residual Plots

A. Residuals—the difference between an observed value of y and the predicted value from the regression line. The formula for residual is:

Residual = observed y – predicted y
$$= y - \hat{y}$$

1. Although the vertical distance from data points and the least squares regression line are as small as possible, the space left between the line and the data points are the residuals.

Example

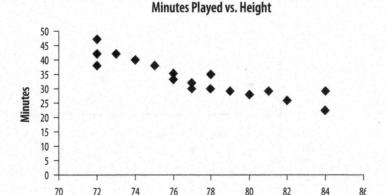

Minutes Played vs. Height

Using the basketball player data, it was found that the observed number of minutes for a player of 78 inches in height was 40 minutes. Determine the residual for this player based on the regression line.

$$\hat{y} = 152.9 - 1.54x$$

$$\hat{y} = 152.9 - 1.54(78) = 32.8 \text{ min}$$

Residual = observed y – predicted y
$$= y - \hat{y}$$
$$40 - 32.8 = 7.2 \text{ min}$$

B. Residual Plot—a scatterplot of the regression residuals against the predicted *y*-values. The residual plot is a "barometer" for how well the regression line fits the data.

1. The residuals of the least square data always sum to zero.

2. In any residual plot, the line indicates the mean of the residuals (since the sum of the residuals is 0, the line *must* be horizontal).

3. The residual plot allows one to see patterns of data in a simpler fashion.

C. Reading a residual plot

1. An unstructured scatter of points around the *y* = 0 line indicates the regression line is a good fit for the data.

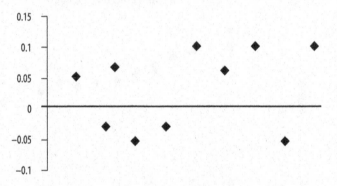

2. A curved scatter of points around the *y* = 0 line indicates the regression line is *not* a good model for the data.

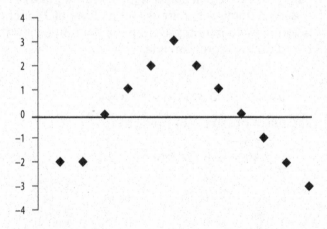

3. A spread around as *x* increases or decreases indicates smaller or larger values of *x* are not accurate.

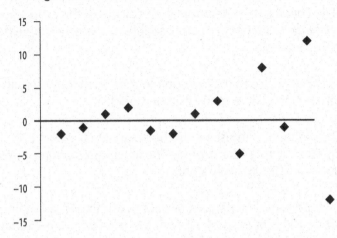

D. Using a calculator for a residual plot

1. The regression line data from the previous example must already be entered and plotted on the calculator.

2. In L_3 enter Y1(L_1) and press ENTER.

3. In L_4 enter the command $L_2 - L_3$ and press enter

4. Turn PLOT 1 OFF.

5. Turn PLOT 2 on and use L_1 as *x* variable and L_4 as *y* variable.

6. Plot the data and press ZoomStat.

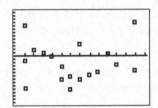

Correlation, least squares regression, and residuals plot are three topics that are usually rolled into one for the AP® test. Look for these concepts to be highly tested in multiple-choice questions or the investigative task question.

VI. Coefficient of Determination (r^2)

A. The coefficient of determination calculates the proportion of the variance (variability) of one variable that is predictable from the other variable.

B. It is a measure that allows one to determine how certain a prediction can be made from a model.

C. The value of r^2 must satisfy the inequality $0 \leq r^2 \leq 1$. The closer its value is to 1, the stronger the linear association between the two variables (usually x and y).

D. r^2 represents the percent of data that is explained by the line of best fit. For example, if $r = 0.88$, then r^2 is approximately 0.77. This means that 77% of the total variation in y can be explained by the linear relationship between x and y in the regression equation. The other 23% of the total variation in y is unexplained.

E. If the regression line passes through every point on the scatterplot, r^2 would equal 1, and thus 100% of the total variation would be explained. Therefore, a good regression line minimizes the sum of the vertical distances from the given points to the line.

VII. Outliers

A. With respect to bivariate data, an outlier is any point that is far removed from the general data pattern of the (least squares) regression line.

B. An outlier in the x direction that exerts a strong influence on the location of the regression line is an influential point.

 Test Tip

You must know the difference between the meanings of r and r^2. The variable r measures the strength of the linear correlation between x and y. The variable r^2 describes what percent of the total data can be explained by the regression line.

Exploring Bivariate Data
(Part 2—Transformations)

How to Transform Data

A. Linear transformations cannot always straighten a curve into a useful plot between two variables.

B. Functions that are non-linear, such as logarithmic, positive, and negative powers, will transform the data accurately.

C. Standard Linear Data can be graphed in the form of

$$\hat{y} = b_0 + b_1 x$$

Example

In chemistry, Boyle's Law is a classic example of how transforming data can lead to a result/graph that allows for better interpretation. In a normal Boyle's Law experiment, the pressure and volume of a gas are measured to determine their relationship. The following is an example of the volume and pressure data. As volume increases, the pressure of a gas decreases. If the data is transformed by plotting 1/pressure, a linear relationship is achieved, allowing one to interpret data more easily.

Pressure (psi)	Volume (mL)
64.3	8.0
44.0	11.7
33.9	15.2
22.9	22.5
14.0	36.8

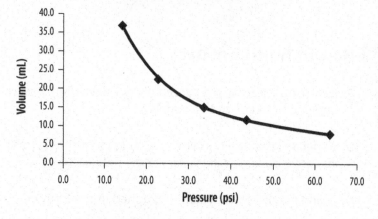

1/Pressure	Volume (mL)
0.0155	8.0
0.0227	11.7
0.0295	15.2
0.0437	22.5
0.0715	36.8

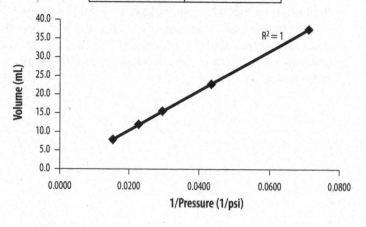

II. Logarithmic Transformations

A. Exponential Growth—increases by a fixed percentage of the previous total in equal time. Examples include populations, technology usage, sports tournaments (decreasing examples), radioactive decay, and monetary investments.

Examples of Plots that are logarithmic:

Year	Cell Phone Users
1986	503
1987	890
1988	1545
1989	2701
1990	4734
1991	8345
1992	14356
1993	25019
1994	45673

Note in the plot below that a linear fit would not accurately depict the data.

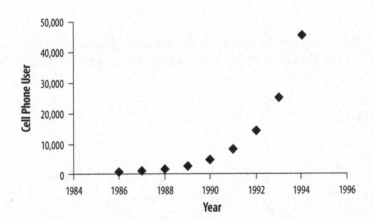

NCAA Tournament

Round	Teams Remaining
1	64
2	32
3	16
4	8
5	4
6	2

Note in the plot below that a linear fit would not accurately depict the data.

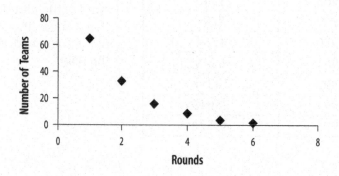

B. Looking more closely at the data on cell phone usage, the following transformation would better describe the relationship period.

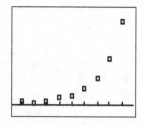
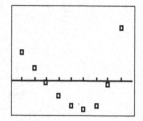

L1	L2	L3	1
1986	503	------	
1987	890		
1988	1545		
1989	2701		
1990	4734		
1991	8345		
1992	14356		

L1(1) = 1986

When a residual plot of the data above is constructed, the curved pattern indicates the linear regression line is not a good fit for the data.

L1	L2	L3 3
1986	6.2206	1.8279
1987	6.7912	1.9156
1988	7.3428	1.9937
1989	7.9014	2.067
1990	8.4625	2.1356
1991	9.0294	2.2005
1992	9.5719	2.2588

L3(1) = 1.827864784...

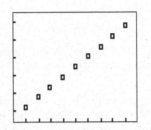
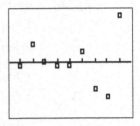

When the data in L2 (cell phone user) is transformed using ln(L2), the curve appears linear and a better relationship can be described. Note that the unstructured scatter of points around the $y = 0$ line indicates the regression line using ln is a good fit for the data.

III. Common Transformations

Transformation	Mathematical Equation
No transformation needed	$\hat{y} = a + bx$
Square Root of x	$\hat{y} = a + b\sqrt{x}$
Reciprocal of x	$\hat{y} = a + b\left(\dfrac{1}{x}\right)$
Log of x	$\hat{y} = a + b\log_{10}(x)$ or $\hat{y} = a + b\ln(x)$
Log of y	$\log(\hat{y}) = a + bx$ or $\ln(\hat{y}) = a + bx$

For the exam, you should know the following definitions and rules about logarithms.

Definitions

1. $\log_a x = N$ means that $a^N = x$.

2. $\log x$ means $\log_{10} x$. All $\log_a$ rules apply for log. When a logarithm is written without a base it means common logarithm.

3. $\ln x$ means $\log_e x$, where e is about 2.718. All $\log_a$ rules apply for ln. When a logarithm is written "ln" it means natural logarithm. Note: ln x is sometimes written Ln x or LN x.

Rules

1. Inverse properties: $\log_a a^x = x$

2. Product: $\log_a(xy) = \log_a x + \log_a y$

3. Quotient: $\log_a\left(\dfrac{x}{y}\right) = \log_a x - \log_a y$

4. Power: $\log_a(x^p) = p \log_a x$

IV. Power Transformations—Both the *x*- and *y*-values Are Transformed According to the Mathematical Statement Below

Power and Name	Mathematical Statement
3 cubic transformation	Both values cubed
2 squared transformation	Both values squared
1 no transformation	Values stay the same
$\dfrac{1}{2}$ square root transformation	Both values square rooted
$\dfrac{1}{3}$ cube root transformation	Both values cube rooted
–1 reciprocal transformation	Both values have reciprocal taken

Graph	Transformation
Curve 1	*x* up and *y* down
Curve 2	*x* down and *y* down
Curve 3	*x* down and *y* up
Curve 4	*x* up and *y* up

V. How to Determine the Correct Type of Transformation

A. Create a scatterplot of the data.

B. These four curved possibilities are most likely to appear on the AP® test after the data is plotted.

C. If a graph looks like curve 1, transform it by moving up the ladder (i.e., from no transformation to a squared or cubic transformation) and down the ladder (i.e., from no transformation to a cube root, log, or reciprocal transformation).

D. All four curves have distinct transformations that can be applied to straighten out the line. At times, only one variable may have to be transformed to achieve a better representation of the data.

Don't worry about an extensive amount of mathematics for transformations on the exam. Most questions about transformations are multiple-choice and will require you to analyze residual plots to pick the correct transformation.

Exploring Categorical Data

I. Two-Way Tables

A. Typically, two-way tables describe two different categorical data sets and have both row data and column data.

Example

A survey was conducted at three separate global companies about the marital status of their employees.

	Marital Status			
	Single	Married	Divorced	Widowed
Company A	53	573	22	2
Company B	35	254	32	4
Company C	17	56	5	0

B. Marginal Distributions—the sum of each categorical set found at the bottom and far right of the table. In the example, it would be the total number of employees in companies A, B, and C and the total of each individual marital status.

	Marital Status				
	Single	Married	Divorced	Widowed	Total
Company A	53	573	22	2	650
Company B	35	254	32	4	325
Company C	17	56	5	0	78
Total	105	883	59	6	1053

C. Marginal Frequencies—percentages of each marginal distribution. Often marginal frequencies are a better method for analyzing categorical data.

1. Each marginal distribution from a two-way table is a distribution for a single categorical value.

2. The most appropriate method for depicting each marginal distribution for a single categorical value is using a bar chart or a pie chart.

Company A

$$\text{Single } \frac{53}{650} = 8.2\% \qquad \text{Married } \frac{573}{650} = 88.2\%$$

$$\text{Divorced } \frac{22}{650} = 3.4\% \qquad \text{Widowed } \frac{2}{650} = 0.3\%$$

Marital Status Total

$$\text{Single } \frac{105}{1053} = 10.0\% \qquad \text{Married } \frac{883}{1053} = 83.9\%$$

$$\text{Divorced } \frac{59}{1053} = 5.6\% \qquad \text{Widowed } \frac{6}{1053} = 0.60\%$$

Note that the sum of the percents is 100.

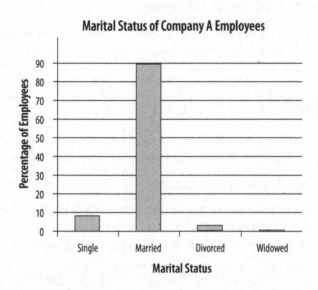

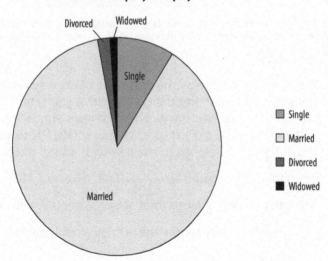

D. Relative/Conditional Frequencies

1. Two-way tables can also display relative frequencies, also known as conditional frequencies, for the entire table, for rows, or for columns.

Example

Refer to the marginal distribution table in 1B on page 73. To illustrate the conditional frequencies for the entry of 53, simply convert the fraction $\dfrac{53}{1053}$ to a decimal, rounded off to the nearest hundredth, 0.05. Each of the other 14 numerical entries, which include row and column totals, is changed similarly. Here is how the entries would appear:*

	Single	**Married**	**Divorced**	**Widowed**	**Total**
Company A	0.05	0.55	0.02	0.00*	0.62
Company B	0.03	0.24	0.03	0.00*	0.30
Company C	0.02	0.05	0.01	0.00	0.08
Total	0.10	0.84	0.06	0.00*	1.00

* Due to rounding, some of these entries are not zero but contribute less than 0.5% to the total. Similar adjustments will also be made for the next two tables, in order that the totals reflect the proper sums.

2. Another way to display the entries of a two-way table is to present conditional frequencies for rows. Each row total changes to 1.00 and the individual entries are converted by dividing them by their respective row totals. For example, the entry of 53 belongs to the row with a total of 650. Therefore, 53 changes to $\dfrac{53}{650} \approx 0.08$. Likewise 105, which is the entry for the column total of single people, will change to $\dfrac{105}{1053} \approx 0.10$. Here is how the entries would appear:

	Single	Married	Divorced	Widowed	Total
Company A	0.08	0.88	0.03	0.01	1.00
Company B	0.11	0.78	0.10	0.01	1.00
Company C	0.22	0.72	0.06	0.00	1.00
Total	0.10	0.84	0.05	0.01	1.00

3. A third way to show conditional frequencies is to present each column total as 1.00. The entry of 53 is changed by dividing it by its corresponding column total (105). So, 53 becomes $\frac{53}{105} \approx 0.51$. Similarly, the entry of 573 is changed to $\frac{573}{883} \approx 0.65$. Here is the entire table of entries:

	Single	Married	Divorced	Widowed	Total
Company A	0.51	0.65	0.37	0.33	0.62
Company B	0.33	0.29	0.54	0.67	0.31
Company C	0.16	0.06	0.09	0.00	0.07
Total	1.00	1.00	1.00	1.00	1.00

Test Tip

Two-way tables require you to calculate percentages on the bottom and far right of the table. Make sure the percentages sum to 100; finding the association between the categorical values should be simple.

II. Simpson's Paradox

A. Based on lurking variables (see glossary) that are categorical in nature.

B. A comparison that is true for several groups can reverse direction when the data is combined to form a single group.

Example

The best example of Simpson's Paradox can be observed in admission rate based on gender. In Table A it seems as if males are more likely to be accepted to College X than females.

Table A

Gender	Applicants	% Admitted
Male	7564	42
Female	5341	34

Table B

Department	Men		Women	
	Applicants	% Admitted	Applicants	% Admitted
W	467	62	55	82
X	324	63	45	68
Y	234	37	515	34
Z	423	33	679	30

When looking at the number of applicants more closely using the categorical value of each department in Table B, it was found that more women applied to departments Y and Z that have low rates of admission when compared to the other departments. The admission rates for these two departments are not statistically different.

Test Tip

You may have to take the data from a two-way table and express it in another form—a bar chart or a pie chart. You may then have to analyze the data further and determine what the lurking variable is—hence Simpson's paradox.

PART III:
SURVEY METHODS

Overview of Methods of Data Collection

I. Population

Population is defined as the group of individuals from whom data is collected.

Example

The entire 25-man roster of a baseball team.

II. Sample

A sample is a portion of a population that is examined and from which data is collected.

Example

The starting pitchers of a 25-man baseball team roster.

III. Census

A census attempts to contact every individual in a population.

Example

A poll of all 450 National Basketball Association players is conducted to determine who is the best referee in the league.

IV. Sampling

A sampling studies a portion of the population to determine data about the whole population.

Example

Local New Jersey high school students are sampled to determine their favorite subject in order to make a determination about all New Jersey high schools.

V. Observational Study

An observational study is one in which the individuals are observed and specific variables of interest are measured. The observer *does not* influence or bias the response of the individuals. A direct cause-and-effect relationship cannot be determined from an observational study.

Examples

A study to estimate the long-term psychological effects on children ages 5–10 after losing a parent to sudden death.

A study commissioned by Major League Baseball to estimate the effects of the number of home runs hit in day games versus night games during an entire baseball season.

A research study determining the rate of the incidence of cancer in persons who followed a Mediterranean diet.

The problem with observational studies is the lurking variable. Other variables could be the true cause of a particular outcome of a study; thus, the researcher's conclusion could be based on incorrect reasoning.

VI. Experiments

In experiments, individuals are subjected to a particular treatment and the responses to the treatment are measured. Experiments show a cause-and-effect relationship between two variables.

Examples

A study is commissioned by oncologists who treated breast cancer patients with the drug tamoxifen to determine whether it is a viable drug.

In 2002, Major League Baseball experimented with baseballs at Coors Field in Colorado by placing them in a humidor to test whether this affected the number of home runs hit at Coors Field.

Yeast cells were grown in various concentrations of glucose to determine which concentration allowed for maximum cellular respiration.

Knowing the difference between an observational study and an experiment has been on every released AP® Statistics test. Observational studies, though not meaningless, are not as informative as experiments and, therefore, are not better for obtaining a cause-and-effect relationship.

VII. Voluntary Response Sampling

Voluntary response sampling refers to individuals who choose to respond to surveys or polls.

Example

Voluntary response sampling is biased because those with strong negative opinions usually respond. Examples include call-in radio or television shows.

VIII. Convenience Sampling

In convenience sampling, individuals are chosen because they are easiest to reach.

Example

Sampling that is done at major public areas such as airports, malls, food stores, or amusement parks.

IX. Bias

Bias refers to sampling methods that favor a certain outcome.

Both voluntary response sampling and convenience sampling lead to biased outcomes.

Biased outcomes can lead to outlier data or conclusions that are not necessarily correct. Included is skewed data that can be observed on graphs, dotplots, or boxplots. Simple Random Samples (Chapter 9) will control for biased outcomes so data is representative of the population.

Planning and Conducting Surveys

I. Characteristics of a Well-Designed and Well-Conducted Survey

A. Be certain that bias is not in the survey (the designer of the survey must have no personal preference or connection when designing the survey).

B. Avoid undercoverage (when groups of the population are left out of the process of choosing the sample).

Example

Performing a survey at an amusement park in the summer might exclude those who are retired or work during the day. The survey would be limited to young people who tend to frequent such venues.

C. Limit nonresponses (when specific individuals of a survey group cannot be contacted or refuse to cooperate with the survey).

Example

Contacts will more likely respond to a phone poll after work at 6pm rather than during the normal workday at 1pm.

D. Response bias (when an interviewer or respondent behavior skews the data).

Example

If a phone poll asks whether respondents ever cheated on their taxes, the answer would probably be a resounding "no." Persons will not admit to swindling the government even if committing some low-level cheating on their taxes.

E. Wording of questions (do not confuse respondents or lead them to a specific answer).

Example

A phone poll asked, "Do you think voting down the school budget and teachers losing their jobs would help increase mathematics scores districtwide?" Obviously this question is leading the respondent to a negative answer, and presents only one side of the issue.

F. The date and time of a poll (a poll conducted in 2010 and used as a data point for 2018 data may bias the data).

Example

In 2012, a poll asked how many families purchased a cell phone for each of their children. With the boom of technological advances in smartphones and the lower cost of basic cell phone data plans, more families in 2018 could afford to buy cell phones for their children.

G. Large random samples will give more detailed and accurate results.

Example

If a high school in suburban Chicago has 2200 students enrolled, then taking a random sample of 220 students about parking privileges will provide better data than sampling one 9th grade class of 30 students.

One multiple-choice AP® test question will require you to know the essentials of good surveying methods. You may have to choose from a list of reasons about why a certain survey is flawed or not flawed.

II. Sampling Methods

A. Simple Random Sample (SRS)—when a subset of individuals is chosen from a larger group. Each person is chosen randomly and by chance, so that each individual has the same probability of being part of the sample. Each individual and their probability of being chosen is independent from one another.

1. An SRS makes use of random digits and a random number table as seen below.

2. Random numbers are sets of digits (i.e., 0, 1, 2, 3, 4, 5, 6, 7, 8, 9) arranged in no real order. This means no individual digit can be predicted from knowledge of any other digit or group of digits.

71349	13410	60667	45454	37977	60408	93116	21551
62803	39045	57204	02729	92711	49322	52931	15547
54258	64681	80298	60004	20887	38236	39305	36100
45713	90316	03393	17278	75621	00593	99120	30096
37168	42509	26487	74553	03797	89507	58935	24092
28623	68144	23023	31828	58531	51863	18751	44645
93520	93779	46118	62544	86707	40777	78566	38641
84975	19415	69212	19819	41441	29691	64940	32637
76430	45050	65749	77094	69617	92048	24755	53190
67885	70685	88843	34368	24351	80962	84571	47186
59340	96320	11938	91643	52527	43318	44386	41182
50795	21955	35032	48918	80703	32232	30759	

Example

A food store wants to check on the sales of its organic food products. So, the manager assigns numbers to 25 organic products starting with 01 for organic bananas and ending with 25 for organic chicken. To avoid bias, the manager chooses 5 items using SRS.

01-bananas	11-red bell peppers	21-yellow peppers
02-grapes	12-kale	22-spinach
03-olives	13-avocado	23-corn
04-lettuce	14-milk	24-chocolate
05-tomatoes	15-orange juice	25-chicken
06-cucumbers	16-BBQ sauce	
07-ground beef	17-kidney beans	
08-hot dogs	18-turkey	
09-pasta	19-ham	
10-soy nuts	20-lemons	

71 34 91 34 <u>10</u> 60 66 74 54 54 37 97 76 <u>04</u>
<u>08</u> 93 <u>11</u> 62 15 51 62 80 33 90 45 57 <u>20</u>

Using a portion of the first line of random numbers, two-digit numbers are represented in order. The underlined two digit numbers correspond to 5 different organic products the manager originally labeled. The sample includes products 04, 08, 10, 11, and 20 or lettuce, hot dogs, soy nuts, red bell peppers, and lemons.

B. Stratified Random Sample—the entire group of individuals is divided into groups called strata (singular stratum). Strata are similar in some fashion (homogeneous group) that is relevant to the study. Within the individual stratums, a distinct single random sample is chosen and those groups are combined to form the full sample.

Example

A high school survey could be divided into grade strata: 9th, 10th, 11th, and 12th. An SRS of individuals can be selected from each grade and surveyed. Therefore, a stratified random sample can give you data on each particular stratum, as well as the population as a whole.

C. Cluster Sampling—the population is divided into clusters, and all individuals in the chosen clusters are selected to be in the group that will be sampled.

Example

A survey gathers student opinions on the most recent Algebra 1 field test in Colorado. First, only high schools that offered the test would be chosen, and afterward, only a cluster of them would be surveyed. From the cluster of high schools, *all* students who took the field test would be surveyed.

D. Multistage Sampling—choosing samples in stages.

Test Tip

Very likely your AP® test will have you differentiate facts about a simple random sample and a stratified random sample.

Planning and Conducting Experiments

I. Characteristics of a Well-Designed and Well-Conducted Experiment

A. Experimental units—the individuals on which the experiment is being conducted.

B. Subjects—when the experimental units are human beings.

C. Treatment—the specific condition that is applied to the experimental units.

D. Placebo effect—the inert treatment that is essential for a controlled experiment.

Example

1. To rate the efficacy of a weight-loss drug, 10 persons are given a weight-loss pill daily, while 10 others receive a placebo pill. Each week, all 20 people are weighed. Those taking the weight-loss pill should lose more weight if the diet drug has true efficacy.

2. Understand that the placebo and the experimental treatment may render the same results, indicating the treatment had no differential effect.

3. Those subjects being treated with the placebo tend to believe it has some effect, because of their trust in the scientific process and the legitimacy of the study.

E. Control Group—the group of subjects that receive a placebo or an alternative treatment.

F. Control (scientific)—when an experiment is conducted for the purpose of determining the effect of a single variable of interest on a particular system, a scientific control is used to minimize the unintended influence of other variables on the same system.

G. Replication—the repetition of an experiment on a large group of subjects.

1. Replication is required to improve the significance of an experimental result. If a treatment is truly effective, the long-term averaging effect of replication will reflect its experimental worth.

2. Replication reduces variability in experimental results, increasing their significance and the confidence level with which a researcher can draw conclusions about an experimental factor.

H. Randomization—using chance to assign subjects to treatment conditions.

1. Statistically Significant Experimental Result—an observed effect or result is unlikely to occur by chance.

Test Tip

An understanding of the scientific method can assist you in designing an effective experiment. If asked on the free-response section to design an experiment, you must use the preceding terms correctly in the experimental design.

II. Randomized Comparative Design

A. Three key parts of a randomized comparative design experiment would be to:

1. control lurking variables by comparing only one or two treatments at a time;

2. replicate the experiment to reduce variation and ensure the efficacy of the results;

3. randomize the design to produce similar groups of subjects so that valid comparisons can be made.

B. Completely Randomized Design—all experimental units are chosen at random for all treatments.

1. A completely randomized design makes use of simple random sample methods to assign groups.

III. Blocking

A. Block—a group of experimental units or subjects that are known before the experiment to be similar. The characteristics of the group are known before the experiment and are controlled by placing experimental units or subjects into similar groups. The hope of blocking is to reduce variations among the experimental units. Gender separation and age separation are common forms of blocking.

IV. Matched Pair Design

A. Compares two treatments.

B. Matched subjects are more common to each other and share characteristics, allowing comparisons with less variation.

V. Double Blind Study

A. Neither the subject nor those who measure the response of the subject know the nature of the experimental treatment.

Blocking will surely be covered on the AP® Statistics exam. The basic concept is to create homogeneous blocks in which variation and the lurking variable are minimized/negated. Within blocks, it is possible to assess the effect of different levels of the factor of interest without worrying about variations due to changes of the block factors. Follow this rule in experimental design:

"Block what you can, randomize what you cannot."

PART IV:

PROBABILITY AND
EXPECTED VALUE

Probability

I. Interpreting Probability, Including Long-Run Frequency Interpretation

A. The key to probability is that the relative frequency of events is unpredictable in the short term, but in the long term has a regular pattern.

Example

If you toss a coin in the air, there is a 50% chance of either heads or tails appearing. If the coin is tossed ten times, the split between heads and tails (or vice versa) could be 70% to 30%. As more and more coin tosses take place, the proportion of heads to tails will eventually get closer to 50%–50%. This is also called the Law of Large Numbers.

B. Probability is also empirical, that is, based on observation.

C. Random individual outcomes are uncertain, but after long repetitious trials, a regular pattern can be inferred.

D. Trials that lead to probabilities must be independent of each other. For example, the experimenter cannot influence chances of randomness, such as stacking a deck of cards with more face cards when playing poker.

E. A model for studying probability is based on the following:

1. Sample space (S)—the set of all possible outcomes. In the coin-tossing example above, it would be heads (H) to tails (T), otherwise written as S = {H,T}.

2. Event—the outcome or set of outcomes of a random phenomenon. Thus, {T} would be the event of tossing tails in the coin example.

3. Probability model—a mathematical description of the random phenomenon in the sample space by assigning probabilities to each event. For example, the probability for heads is 0.5, for tails, also 0.5.

Example of Multiplication Principle

You toss a coin and pick one of four queen face cards. The following example uses a tree diagram to illustrate all the possible outcomes.

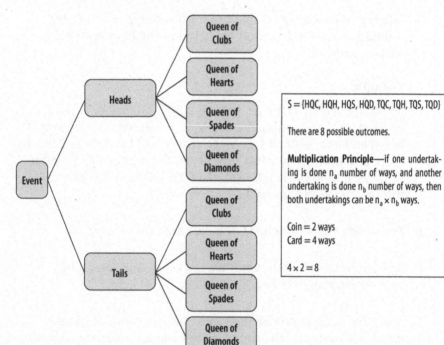

S = {HQC, HQH, HQS, HQD, TQC, TQH, TQS, TQD}

There are 8 possible outcomes.

Multiplication Principle—if one undertaking is done n_a number of ways, and another undertaking is done n_b number of ways, then both undertakings can be $n_a \times n_b$ ways.

Coin = 2 ways
Card = 4 ways

$4 \times 2 = 8$

F. Sampling with replacement—placing the original items from a sample back into the sample to be drawn again. For example, if you choose one card from a deck of 52 cards and place that card back into the sample, you still have 52 cards to choose from on the next draw.

G. Sampling without replacement—not placing the original items from a sample back into the sample to be drawn again. For example, if you choose one card from a deck of 52 cards, and do not place that card back into the sample, you have 51 cards to choose from on the next draw.

II. Rules of Probability

A. Any probability is a number between 0 and 1.

1. For example, the probability P of event X can be expressed as $0 \le P(X) \le 1$.

B. The sum of all probabilities must be equal to 1.

1. $P(S) = 1$

C. If two events have no outcomes in common, the probability that one or the other occurs is the sum of their individual probabilities.

1. If one event occurs in 20% of all trials and another event occurs in 70% of all trials, and the two events can never occur together, then one or the other occurs 90% on all trials.

2. The two events are said to be *mutually exclusive* or *disjoint* (cannot occur at the same time) or $P(A \text{ or } B) = P(A) + P(B)$.

3. Another way to express mutually exclusive events is $P(A \cup B) = P(A) + P(B)$. The word "or" means union.

Example

Below is an example using a Venn Diagram.

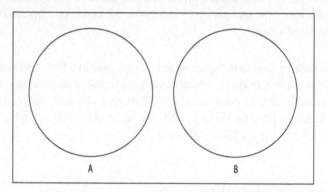

A   B

If two events A and B are independent, that is, the outcome of one event does not affect the outcome of the other event, the probability of both events is the product of their individual probabilities. The multiplication rule only works for events that are *independent* and not for mutually exclusive events.

1. $P(A \text{ and } B) = P(A)P(B)$

2. Another way to express independent events is $P(A \cap B) = P(A)P(B)$. The word "and" means intersection.

3. If a fair coin is tossed in the air, what is the probability of the first toss being heads and the second toss being tails?

First Toss   0.5 or $\frac{1}{2}$ probability of heads

Second Toss   0.5 or $\frac{1}{2}$ probability of tails

Total Probability $\frac{1}{2} \times \frac{1}{2} = \frac{1}{4}$

Example

Below is an example using a Venn Diagram

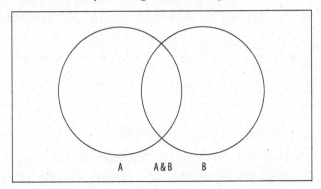

A   A & B   B

D. The probability that an event does not occur is 1 minus the probability that the event does occur. This is otherwise known as the complement rule.

1. If an event occurs 25% of the time, then it does not occur 75% of the time: 0.25 + 0.75 = 1.00.

2. Another way to state this rule is: if A^c represents the complement of any event, then $P(A^c) = 1 - P(A)$.

The union of two events corresponds to the word "or" while the intersection of events corresponds to the word "and." For example, if Set A contains the numbers {1, 2, 3} and Set B the numbers {2, 3, 5}

$$A \cup B = \{1, 2, 3, 5\}$$

$$A \cap B = \{2, 3\}$$

Union means the list of elements in either set, while intersection means the elements that occur in both sets.

If two events are mutually exclusive, then $P(A \cap B) = 0$, meaning A and B cannot occur together.

Examples Using Rules of Probability

1. Below is a table of the heights of college basketball players and the probability of being drafted to play professionally.

Height	6'0" to 6'3"	6'4" to 6'6"	6'7" to 6'10"	6'11" to 7'2"	> 7'2"
Probability of being drafted	0.21	0.29	0.36	0.10	0.04

a. Describe the data in the table.

First, the sum of the probabilities equals 1. The table indicates the chances of a college basketball player being drafted *only* based on height.

b. What percent of the draftees are greater than 6'3"?

Use the complement rule $P(A^c) = 1 - P(A)$.

A = the probability of being drafted below 6'3" or 0.21

A^c = the probability of being drafted above 6'3" or $1 - 0.21 = 0.79$

c. If a player is taller than 6'6", what are his chances of being drafted?

Use the rule: $P(A \text{ or } B) = P(A) + P(B)$

$P(A)$ = 6'7" to 6'10" or 0.36

$P(B)$ = 6'11" to 7'2" or 0.10

$P(C)$ = > 7'2" or 0.04

$P(A \text{ or } B \text{ or } C) = P(A) + P(B) + P(C) = 0.36 + 0.10 + 0.04 = 0.50$

These events are mutually exclusive. You cannot have the probability of two heights at the same time. Also note that the rule incorporated a third probability of C.

More Complex Probability Rules

A. Addition rule for disjoint events *A*, *B*, and *C* (this rules extends to any number of disjoint events):

$$P(A \text{ or } B \text{ or } C) = P(A) + P(B) + P(C)$$

B. General Addition Rule for the Union of 2 Events:

If events A and B are *not disjoint*, they can occur simultaneously.

$$P(A \text{ or } B) = P(A) + P(B) = P(A \text{ and } B)$$

or

$$P(A \cup B) = P(A) + P(B) = P(A \cap B)$$

Example

The probability that Matt makes the varsity football team is 0.4. The probability that Ryan makes the varsity football team is 0.7. The probability of both of them making the varsity football team is 0.2. Find the following:

A) At least one of them makes the varsity team.

$$P(\text{at least one makes the team}) = P(A) + P(B) - P(A \cap B)$$

$$P(0.4) + P(0.7) - P(0.2) = 0.9$$

B) Draw both of them making the varsity team using a Venn Diagram.

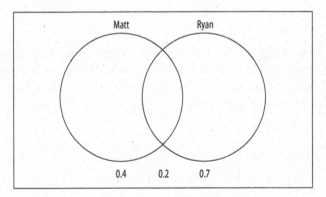

C) Neither of them makes the varsity team.

The complement rule indicates that $1 - 0.9 = 0.1$ that neither of them makes the varsity team. To indicate neither on the varsity team, place 0.1 in the right bottom corner of the Venn Diagram.

IV. Conditional Probability

A. The probability assigned to an event can change based on a separate event taking place.

$$P(A \cap B) = P(A)P(B|A)$$

$P(B|A) =$ is the conditional probability that B occurs after given information that A occurs.

Example

There are 9 balls in a bag. 7 are green and 2 are blue. Find the probability of choosing a green ball followed by a blue ball without replacement.

$$P(G) = \frac{7}{9}$$

$$P(B|G) = \frac{2}{8} = \frac{1}{4}$$

$$P(G \cap B) = \frac{7}{9} \times \frac{1}{4} = \frac{7}{36}$$

Example

The probability that individuals in a sporting goods store are football fans is 0.80. If 0.15 of those football fans are New Orleans Saints fans, what is the percent of football fans in the store who are Saints fans?

$$P(F) = 0.8$$

$$P(S|F) = 0.15$$

$$P(F \cap S) = 0.8 \times 0.15 = 0.12 = 12\%$$

The conditional probability formula can be rearranged to solve for conditional probability directly.

$P(B \mid A) = \dfrac{P(A \cap B)}{P(A)}$ (this is a re-arrangement of the equation under A on pg. 104).

Example

75% of people in community X own a smartphone. 20% have both a smartphone and a laptop computer. What is the probability that a person chosen at random from community X will have a laptop computer, given that this person already owns a smartphone?

$P(lap \mid sp) = \dfrac{P(sp \cap lap)}{P(sp)} = \dfrac{0.2}{0.75} = 0.2\overline{6}$

V. Discrete Random Variable and Their Probability Distributions Including Binomial and Geometric

A. Random Variable (RV)—a variable whose value is a numerical outcome of a random phenomenon.

B. Random variables are always denoted with a capital letter such as X.

Example

A single die is rolled in which the outcomes can be 1–6. Let X (the RV) be equal to the number of odd numbers that could be outcomes. In this case, the RV $X = 3$ for the number 1, 3, or 5.

C. Discrete Random Variable (DRV)—a variable that has a countable number of outcomes. Using the example of a single die being rolled, X is the DRV for all the outcomes with the probability distributions that follow.

Value of X	1	2	3	4	5	6
Probability	$\frac{1}{6}$	$\frac{1}{6}$	$\frac{1}{6}$	$\frac{1}{6}$	$\frac{1}{6}$	$\frac{1}{6}$

➤ Note all probabilities are between 0 and 1.

➤ The sum of all probabilities is equal to 1.

➤ The probability of a 4 or greater: $P(X \geq 4) = \frac{1}{6} + \frac{1}{6} + \frac{1}{6} = \frac{1}{2}$.

Example

What is the probability distribution of the discrete random variable X that counts the number of girls a woman could give birth to if she has 4 children?

B = Boy and G = Girl

BBBB	BBBG BBGB BGBB GBBB	GBBG GBGB GGBB BGGB BGBG BBGG	GGBG GBGG BGGG GGGB	GGGG
$X = 0$	$X = 1$	$X = 2$	$X = 3$	$X = 4$
$P(X = 0) =$ $\frac{1}{16}$	$P(X = 1) =$ $\frac{1}{4}$	$P(X = 2) =$ $\frac{6}{16} = \frac{3}{8}$	$P(X = 3) =$ $\frac{1}{4}$	$P(X = 4) =$ $\frac{1}{16}$

Sample Questions

1. The probability of at least 1 girl?

$$P(X \geq 1) = 1 - P(X = 0) = 1 - \frac{1}{16} = \frac{15}{16}$$

2. The probability of 3 or more girls?

$$P(X \geq 3) = P(X = 3) + P(X = 4) = \frac{1}{4} + \frac{1}{16} = \frac{5}{16}$$

VI. Binomial Distributions

A. Each observation is in one of two categories: successes or failure.

B. There is a fixed number of observations.

C. All observations are independent.

D. The probability of success, called p, is the same for each observation.

E. X = the random variable of the number of successes.

F. The probability distribution is called the binomial distribution.

$X = B(n, p)$

n = # of observations

p = probability

G. The Binomial Coefficient—the number of ways of arranging k success among n observations.

$$\binom{n}{k} = \frac{n!}{k!(n-k)!}$$

H. Binomial Probability—if X is a binomial distribution with n observations and probability p of success on each observation, and k is any one of the possible values of X then

$$P(x = k) = \binom{n}{k} p^k (1-p)^{n-k}$$

Example

An airline company is testing brake pads that are used to slow commercial airplanes upon runway touchdown. The company has chosen 15 brake pads from a sample of 15,000. Only 5% are deemed "insufficient for usage." What is the probability that no more than 2 brake pads will be deemed insufficient for usage?

$n = 15$

$p = 0.05$

$P(X \leq 2) = P(X = 0) + P(X = 1) + P(X = 2)$

$P(X = 0)$

$\binom{15}{0}(0.05)^0(0.95)^{15} = 0.46$

$P(X = 1)$

$\binom{15}{1}(0.05)^1(0.95)^{14} = 0.37$

$P(X = 2)$

$\binom{15}{2}(0.05)^2(0.95)^{13} = 0.13$

Thus, $P(X \leq 2) = 0.46 + 0.37 + 0.13 = 0.96$

1. Mean and Standard Deviation of a Binomial Random Variable

   Mean $= \mu = np$

   Std Dev $= \sigma = \sqrt{np(1-p)}$

Example

Using the previous example of airplane brakes:

   Mean $= \mu = np = 15 \times 5 = 0.75$

   Std Dev $= \sigma = \sqrt{np(1-p)} = \sqrt{0.75(1-0.05)} = 0.84$

VII. Geometric Distribution

A. Each observation falls into one of two categories: success or failure.

B. The observations are independent.

C. The probability of success, called p, is the same for each observation.

D. The variable of interest is the number of trials required to obtain the first success.

E. Formula for geometric probability.

$$P(X = n) = (1 - p)^{n-1}p$$

p = probability of success

$1 - p$ = probability of failure

F. Mean and Standard Deviation of a Geometric Random Variable

$$\text{Mean} = \mu = \frac{1}{p}$$

$$\text{Variance} = \frac{(1-p)}{p^2}$$

G. The probability that it takes more than n trials to encounter the first success.

$$P(X > n) = (1 - p)^n$$

Test Tip

Don't get overwhelmed with memorizing equations such as those for binomial distribution. All of these equations will be provided in the reference section of the test. Be smart . . . See Appendix A at the back of this book or go to the College Board website and download the reference sheet for the AP® Statistics exam. If the equations do not look familiar to you, spend some time studying them.

Combining Independent Random Variables

I. The Mean of a Random Variable

A. The mean of a probability distribution describes the long-run average outcome, and not the common arithmetic mean described earlier.

B. The symbol for the mean of a probability distribution (X) is mu (μ_x).

C. The mean of any discrete random variable is a weighted average in which each outcome is multiplied by its probability.

Value of X	x_1	x_2	x_3	.....	x_n
Probability	p_1	p_2	p_3	.....	p_n

$$\mu_x = x_1 p_1 + x_2 p_2 + x_3 p_3 + ... + x_n p_n$$

Example

The probability of selling X number of the same car per day over a 5-day period at a local dealership is depicted in the table below. If each car costs \$27,000, what is the expected daily total dollar amount taken in by the dealer from the sale of the cars?

Value of X	0	1	2	3	4
Probability	0.05	0.50	0.35	0.07	0.03

$$\mu_x = x_1 p_1 + x_2 p_2 + x_3 p_3 + ... + x_n p_n$$
$$\mu_x = 0(0.05) + 1(0.50) + 2(0.35) + 3(0.07) + 4(0.03) = 1.53 \text{ cars}$$

On average the dealer sells 1.53 cars per day which results in a daily dollar amount of 1.53 × $27,000 = $41,310.

II. Variance of a Discrete Random Variable

A. Since the mean or measure of center is different for a discrete random variable, the variance and standard deviation (measure of spread from the mean) must also be different.

Value of X	x_1	x_2	x_3	.....	x_n
Probability	p_1	p_2	p_3	.....	p_n

The variance of X is defined as:

$$\sigma^2 = (x_1 - \mu_x)^2 p_1 + (x_2 - \mu_x)^2 p_2 + (x_3 - \mu_x)^2 p_3 + ... + (x_n - \mu_x)^2 p_n$$

The standard deviation of $X(\sigma_x)$ is the square root of the variance.

Example

Based on the car example above, calculate the standard deviation.

$$\sigma^2 = (x_1 - \mu_x)^2 p_1 + (x_2 - \mu_x)^2 p_2 + (x_3 - \mu_x)^2 p_3 + ... + (x_n - \mu_x)^2 p_n$$

$$\sigma^2_x = (0 - 1.53)^2 0.05 + (1 - 1.53)^2 0.50 + (2 - 1.53)^2 0.35 + (3 - 1.53)^2 0.07 + (4 - 1.53)^2 0.03 = 0.67$$

$$\sigma_x = 0.82$$

B. Two Independent Random Variables

1. Consider another discrete random variable Y, as shown in the following table.

Value of Y	5	6	7	8
Probability	0.15	0.25	0.30	0.30

The mean $\mu_Y = (5)(0.15) + (6)(0.25) + (7)(0.30) + (8)(0.30) = 6.75$.

The standard deviation is calculated as follows:

$$\sigma^2_y = (5 - 6.75)^2(0.15) + (6 - 6.75)^2(0.25) + (7 - 6.75)^2(0.30) + (8 - 6.75)^2(0.30) = 1.0875.$$

Then $\sigma_Y = \sqrt{1.0875} \approx 1.04$.

Referring to the numerical example that uses the variable X, the mean of the sum of the two (independent) random variables X and Y, denoted as μ_{X+Y}, is the sum of the mean for each of X and Y. Thus, $\mu_{X+Y} = 1.53 + 1.04 = 2.57$. The standard deviation of the sum of X and Y, denoted as σ_{X+Y}, is the square root of the sum of the individual variance of each of X and Y. Thus, $\sigma_{X+Y} = \sqrt{\sigma^2_X + \sigma^2_Y} = \sqrt{0.67 + 1.0875} \approx 1.326$.

The mean of the difference of X and Y, denoted as μ_{X-Y}, is simply the difference of the mean of each of X and Y. Thus, $\mu_{X-Y} = 1.53 - 6.75 = -5.22$.

The standard deviation of the difference of X and Y, denoted as σ_{X-Y} is calculated the same way as the standard deviation of the sum of X and Y. So, $\sigma_{X-Y} = \sigma_{X+Y} = 1.326$.

(The proof is beyond the scope of this book.)

Test Tip

Expect at least one question on the AP® exam that will require you to find the mean of a random variable based on probability.

PART V:

DISTRIBUTIONS OF DATA

The Normal Distribution

I. Properties of a Normal Distribution

A. All normal distributions have the same shape—bell-shaped and symmetric.

B. The mean and median are located at the center of the symmetric or normal distribution.

C. The standard deviation determines the shape of the normal distribution.

D. The normal distribution can illustrate many types of data including test scores and sampling procedures.

E. The 68–95–99.7 Empirical Rule

1. Approximately 68% of the observations in a normal distribution fall within one standard deviation of the mean.

2. About 95% of the observations in a normal distribution fall within two standard deviations of the mean.

3. Approximately 99.7% of the observations in a normal distribution fall within three standard deviations of the mean.

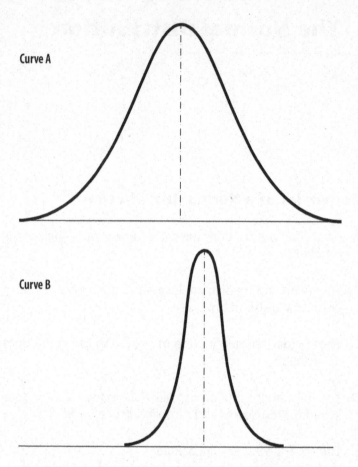

The two examples above indicate two normal distributions with the dashed lines reflecting the means for each set of data.

Curve A is more spread resulting in a larger standard deviation than curve B.

As a result, when the standard deviation is large, the curve is wide. When the standard deviation is small, the curve is narrow.

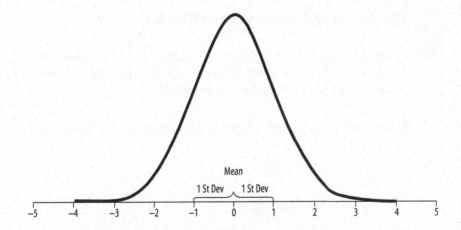

The graph above depicts a normal bell-shaped curve with a mean of 0 and a standard deviation of 1. Notice that the standard deviations can be to the left or right of the mean (which is at the center).

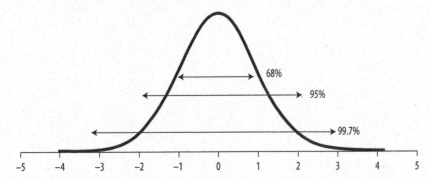

The graph above depicts the 68–95–99.7 Empirical Rule. Almost the entire area under the curve is covered within three standard deviations from the mean.

Test Tip

A standard multiple-choice question will ask you the basic properties of a normal distribution and the properties of curves that represent normal data.

The Standard Normal Distribution

A. The standard normal distribution is the normal distribution with a mean of 0 and a standard deviation of 1. By convention, this is written as N(0,1) or N(mean, standard deviation).

B. To standardize a normal distribution, use the *z*-score formula:

$$z = \frac{x - mean}{s}$$

s = standard deviation

x = original observation

C. A standard normal table provides areas under a standard normal curve. (This table will be provided on the AP® test.)

Standard Normal Table from $z = -4$ to 4

z	0	0.01	0.02	0.03	0.04	0.05	0.06	0.07	0.08	0.09
−4.0	0.0000	0.0000	0.0000	0.0000	0.0000	0.0000	0.0000	0.0000	0.0000	0.0000
−3.9	0.0001	0.0001	0.0000	0.0000	0.0000	0.0000	0.0000	0.0000	0.0000	0.0000
−3.8	0.0001	0.0001	0.0001	0.0001	0.0001	0.0001	0.0001	0.0001	0.0001	0.0001
−3.7	0.0001	0.0001	0.0001	0.0001	0.0001	0.0001	0.0001	0.0001	0.0001	0.0001
−3.6	0.0002	0.0002	0.0002	0.0001	0.0001	0.0001	0.0001	0.0001	0.0001	0.0001
−3.5	0.0002	0.0002	0.0002	0.0002	0.0002	0.0002	0.0002	0.0002	0.0002	0.0002
−3.4	0.0003	0.0003	0.0003	0.0003	0.0003	0.0003	0.0003	0.0003	0.0003	0.0002
−3.3	0.0005	0.0005	0.0005	0.0004	0.0004	0.0004	0.0004	0.0004	0.0004	0.0004
−3.2	0.0007	0.0007	0.0006	0.0006	0.0006	0.0006	0.0006	0.0005	0.0005	0.0005
−3.1	0.0010	0.0009	0.0009	0.0009	0.0008	0.0008	0.0008	0.0008	0.0007	0.0007
−3.0	0.0014	0.0013	0.0013	0.0012	0.0012	0.0011	0.0011	0.0011	0.0010	0.0010
−2.9	0.0019	0.0018	0.0018	0.0017	0.0016	0.0016	0.0015	0.0015	0.0014	0.0014
−2.8	0.0026	0.0025	0.0024	0.0023	0.0023	0.0022	0.0021	0.0021	0.0020	0.0019
−2.7	0.0035	0.0034	0.0033	0.0032	0.0031	0.0030	0.0029	0.0028	0.0027	0.0026
−2.6	0.0047	0.0045	0.0044	0.0043	0.0042	0.0040	0.0039	0.0038	0.0037	0.0036
−2.5	0.0062	0.0060	0.0059	0.0057	0.0055	0.0054	0.0052	0.0051	0.0049	0.0048
−2.4	0.0082	0.0080	0.0078	0.0076	0.0073	0.0071	0.0070	0.0068	0.0066	0.0064
−2.3	0.0107	0.0104	0.0102	0.0099	0.0096	0.0094	0.0091	0.0089	0.0087	0.0084
−2.2	0.0139	0.0136	0.0132	0.0129	0.0126	0.0122	0.0119	0.0116	0.0113	0.0110
−2.1	0.0179	0.0174	0.0170	0.0166	0.0162	0.0158	0.0154	0.0150	0.0146	0.0143
−2.0	0.0228	0.0222	0.0217	0.0212	0.0207	0.0202	0.0197	0.0192	0.0188	0.0183
−1.9	0.0287	0.0281	0.0274	0.0268	0.0262	0.0256	0.0250	0.0244	0.0239	0.0233
−1.8	0.0359	0.0352	0.0344	0.0336	0.0329	0.0322	0.0314	0.0307	0.0301	0.0294
−1.7	0.0446	0.0436	0.0427	0.0418	0.0409	0.0401	0.0392	0.0384	0.0375	0.0367
−1.6	0.0548	0.0537	0.0526	0.0516	0.0505	0.0495	0.0485	0.0475	0.0465	0.0455
−1.5	0.0668	0.0655	0.0643	0.0630	0.0618	0.0606	0.0594	0.0582	0.0571	0.0559
−1.4	0.0808	0.0793	0.0778	0.0764	0.0749	0.0735	0.0721	0.0708	0.0694	0.0681
−1.3	0.0968	0.0951	0.0934	0.0918	0.0901	0.0885	0.0869	0.0853	0.0838	0.0823
−1.2	0.1151	0.1131	0.1112	0.1094	0.1075	0.1057	0.1038	0.1020	0.1003	0.0985
−1.1	0.1357	0.1335	0.1314	0.1292	0.1271	0.1251	0.1230	0.1210	0.1190	0.1170
−1.0	0.1587	0.1563	0.1539	0.1515	0.1492	0.1469	0.1446	0.1423	0.1401	0.1379
−0.9	0.1841	0.1814	0.1788	0.1762	0.1736	0.1711	0.1685	0.1660	0.1635	0.1611
−0.8	0.2119	0.2090	0.2061	0.2033	0.2005	0.1977	0.1949	0.1922	0.1894	0.1867
−0.7	0.2420	0.2389	0.2358	0.2327	0.2297	0.2266	0.2236	0.2207	0.2177	0.2148
−0.6	0.2743	0.2709	0.2676	0.2643	0.2611	0.2578	0.2546	0.2514	0.2483	0.2451
−0.5	0.3085	0.3050	0.3015	0.2981	0.2946	0.2912	0.2877	0.2843	0.2810	0.2776
−0.4	0.3446	0.3409	0.3372	0.3336	0.3300	0.3264	0.3228	0.3192	0.3156	0.3121
−0.3	0.3821	0.3783	0.3745	0.3707	0.3669	0.3632	0.3594	0.3557	0.3520	0.3483
−0.2	0.4207	0.4168	0.4129	0.4090	0.4052	0.4013	0.3974	0.3936	0.3897	0.3859
−0.1	0.4602	0.4562	0.4522	0.4483	0.4443	0.4404	0.4364	0.4325	0.4286	0.4247
0.0	0.5000	0.4960	0.4920	0.4880	0.4840	0.4801	0.4761	0.4721	0.4681	0.4641

Standard Normal Table from $z = -4$ to 4 *(continued)*

z	0	0.01	0.02	0.03	0.04	0.05	0.06	0.07	0.08	0.09
0.0	0.5	0.504	0.508	0.512	0.516	0.5199	0.5239	0.5279	0.5319	0.5359
0.1	0.5398	0.5438	0.5478	0.5517	0.5557	0.5596	0.5636	0.5675	0.5714	0.5753
0.2	0.5793	0.5832	0.5871	0.591	0.5948	0.5987	0.6026	0.6064	0.6103	0.6141
0.3	0.6179	0.6217	0.6255	0.6293	0.6331	0.6368	0.6406	0.6443	0.648	0.6517
0.4	0.6554	0.6591	0.6628	0.6664	0.67	0.6736	0.6772	0.6808	0.6844	0.6879
0.5	0.6915	0.695	0.6985	0.7019	0.7054	0.7088	0.7123	0.7157	0.719	0.7224
0.6	0.7257	0.7291	0.7324	0.7357	0.7389	0.7422	0.7454	0.7486	0.7517	0.7549
0.7	0.758	0.7611	0.7642	0.7673	0.7704	0.7734	0.7764	0.7794	0.7823	0.7852
0.8	0.7881	0.791	0.7939	0.7967	0.7995	0.8023	0.8051	0.8078	0.8106	0.8133
0.9	0.8159	0.8186	0.8212	0.8238	0.8264	0.8289	0.8315	0.834	0.8365	0.8389
1.0	0.8413	0.8438	0.8461	0.8485	0.8508	0.8531	0.8554	0.8577	0.8599	0.8621
1.1	0.8643	0.8665	0.8686	0.8708	0.8729	0.8749	0.877	0.879	0.881	0.883
1.2	0.8849	0.8869	0.8888	0.8907	0.8925	0.8944	0.8962	0.898	0.8997	0.9015
1.3	0.9032	0.9049	0.9066	0.9082	0.9099	0.9115	0.9131	0.9147	0.9162	0.9177
1.4	0.9192	0.9207	0.9222	0.9236	0.9251	0.9265	0.9279	0.9292	0.9306	0.9319
1.5	0.9332	0.9345	0.9357	0.937	0.9382	0.9394	0.9406	0.9418	0.9429	0.9441
1.6	0.9452	0.9463	0.9474	0.9484	0.9495	0.9505	0.9515	0.9525	0.9535	0.9545
1.7	0.9554	0.9564	0.9573	0.9582	0.9591	0.9599	0.9608	0.9616	0.9625	0.9633
1.8	0.9641	0.9649	0.9656	0.9664	0.9671	0.9678	0.9686	0.9693	0.9699	0.9706
1.9	0.9713	0.9719	0.9726	0.9732	0.9738	0.9744	0.975	0.9756	0.9761	0.9767
2.0	0.9772	0.9778	0.9783	0.9788	0.9793	0.9798	0.9803	0.9808	0.9812	0.9817
2.1	0.9821	0.9826	0.983	0.9834	0.9838	0.9842	0.9846	0.985	0.9854	0.9857
2.2	0.9861	0.9864	0.9868	0.9871	0.9875	0.9878	0.9881	0.9884	0.9887	0.989
2.3	0.9893	0.9896	0.9898	0.9901	0.9904	0.9906	0.9909	0.9911	0.9913	0.9916
2.4	0.9918	0.992	0.9922	0.9925	0.9927	0.9929	0.9931	0.9932	0.9934	0.9936
2.5	0.9938	0.994	0.9941	0.9943	0.9945	0.9946	0.9948	0.9949	0.9951	0.9952
2.6	0.9953	0.9955	0.9956	0.9957	0.9959	0.996	0.9961	0.9962	0.9963	0.9964
2.7	0.9965	0.9966	0.9967	0.9968	0.9969	0.997	0.9971	0.9972	0.9973	0.9974
2.8	0.9974	0.9975	0.9976	0.9977	0.9977	0.9978	0.9979	0.9979	0.998	0.9981
2.9	0.9981	0.9982	0.9982	0.9983	0.9984	0.9984	0.9985	0.9985	0.9986	0.9986
3.0	0.9987	0.9987	0.9987	0.9988	0.9988	0.9989	0.9989	0.9989	0.999	0.999

Examples

1. Find the proportion of observations in a standard normal distribution that are less than 1.75.

   Step 1: Locate 1.75 on the standard normal table. Use the left column for 1.7 and the top row for 0.05. The entry in the correct cell is 0.9599 for a $z = 1.75$.

   Step 2: Determine whether the problem is asking you to determine the area from the left or from the right. In the given problem, less than 1.75 would indicate that you are to determine the area from the left.

   Step 3: Draw the normal curve and use an arrow depicting whether the area that is left or right of the z-score. See the graph below.

   Answer: 0.9599 observations are to the left of a $z = 1.75$.

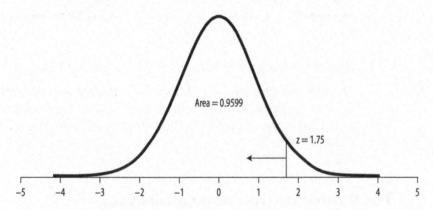

2. Find the proportion of observations in a standard normal distribution that are greater than –2.18.

   Step 1: Locate –2.18 on the standard normal table. Use the left column for –2.0 and the top row for –0.08. The entry in the correct cell is 0.0146 for a $z = 1.75$.

   Step 2: Establish whether the problem is asking you to determine the area from the left or the right. In the given problem, greater than –2.18 would indicate to the right.

Step 3: Draw the normal curve and use an arrow depicting whether the area is left or right of the *z*-score. See the graph below.

Answer: Since we are looking for values to the right of –2.18, subtract 0.0146 from 1: 1 – 0.0146 = 0.9854

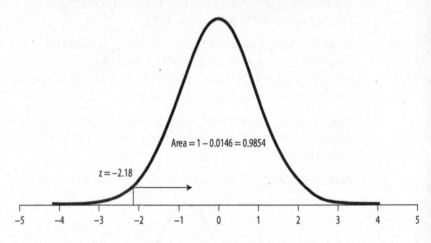

3. Find the *z*-score for an area with 0.45 to the right of it.

To solve this problem, recall that the probabilities must add up to 1. Look for 1 – 0.45 = 0.55. In the table, a probability of 0.55 is closest to 0.5517. The corresponding *z*-score is 0.13.

III. The Normal Distribution Calculations

Important steps:

A. State the problem in terms of the variable *x*.

B. Draw a picture of the distribution.

C. Standardize to find the *z*-score.

D. Use the standard normal table and make a conclusion.

Examples

1. The life expectancy of a laptop computer is distributed normally with a mean of 1000 days and a standard deviation of 110. What is the probability that a laptop computer will last for greater than 1275 days?

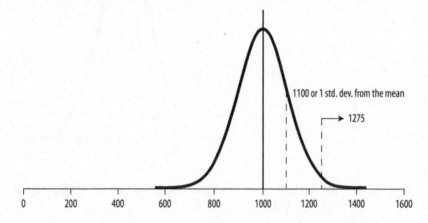

The variable *x* for laptop computers has a distribution of N(1000,110).

Formula for *z*-score: $z = \dfrac{x - mean}{s} = \dfrac{1275 - 1000}{110} = 2.5$

The proportion of observations less than 2.5 is 0.9938 or about 99.38% of laptops will last for less than 1275 days. Only 0.62% will survive more than 1275 days.

2. What is the percent of laptop computers that survive between 670 days and 1050 days?

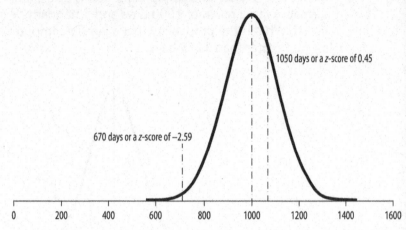

The variable x or laptop computers has a distribution of N(1000,110).

Formula for z-score: $z = \dfrac{x - mean}{s} = \dfrac{670 - 1000}{110} = -3.0$

Formula for z-score: $z = \dfrac{x - mean}{s} = \dfrac{1050 - 1000}{110} = 0.45$

The proportion of observations that have a z-score of −3.0 is 0.0013 or 0.13%. This means that only 0.13% of computers will survive up to 670 days.

The proportion of observations that have a z-score of 0.45 is 0.6736 or 67.36%. This means that 67.36% of computers will survive up to 1050 days.

Therefore, 67.36% − 0.13% = 67.23% of computers will survive between 670 days and 1050 days.

3. In 2001, the average ACT score was a 21 with a standard deviation of 4.7. If a student scored in the top 20%, what was his score?

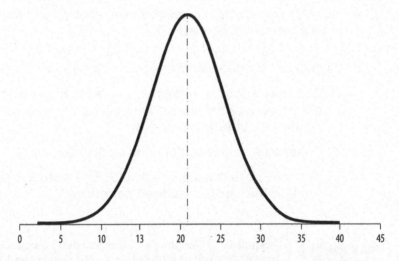

The variable x or ACT scores has a distribution of N(21,4.7).

Using the standard normal table, find the entry in the table closest to 0.80. In the table the entry corresponds to a z-score of 0.84.

Work backwards using the z-score formula.

Formula for z-score:

$$z = \frac{x - mean}{s}$$

$$0.84 = \frac{x - 21}{4.7}$$

$$x = 25$$

A student must have scored a 25 to be in the top 20% of ACT scores in the year 2001.

IV. The Normal Distribution as a Model for Measurements

A. Not all measurements can be considered normal and thus need to be inspected more closely.

B. Steps to test whether data is normal or not normal:

1. Construct a dotplot, histogram, or stem-leaf plot of the data. If the plot has a bell-shaped symmetrical shape, there is a good chance the data is normal.

2. Construct a normality plot using a calculator.

   i. If the points on a normality plot lie close to a straight line, the data is considered normal.

Test Tip

Undoubtedly, calculating a z-score and using a standard normal table will be included on the test. Sample problems in this chapter reflect the various types of questions that may be asked.

Sampling Distributions

I. Sampling Distributions

A. Parameter—a number that describes a population.

B. Statistic—a number that can be determined from sample data without making use of any unknown parameters.

C. Statistics come from samples, but parameters come from populations.

Example

The mean number of cars per household based on a survey of 12,307 households is 2.4. The number 2.4 is a statistic, while the parameter is the mean number of cars for all households.

D. The mean of a population is given the notation mu (μ).

E. The mean of a sample is given the notation ($\overline{x}$).

F. Sampling variability—the value of a statistic will depend on the variation of the random sample.

G. Population Proportion (p) versus Sample Proportion ($\hat{p}$). A sample proportion is used to estimate the population proportion.

Example

A sample of 459 people were surveyed and asked whether they had graduated college. 250 people answered that they did graduate from college.

$$\hat{p} = \frac{250}{459} = 0.54$$

➤ The statistic is 0.54.

➤ The parameter is the total number of college graduates.

➤ The sample proportion of 0.54 is used to estimate the parameter, which is the population proportion.

H. Sampling Distribution—distribution of a given statistic based on a random sample of size n

I. Bias of a Statistic—concerns the center of the sampling distribution and how trustworthy data from a sample proportion ($\hat{p}$) is to estimate a population proportion (p)

J. A statistic is said to be unbiased if the mean sampling distribution of the parameter being measured is equal to the true value being measured.

K. The variability of a statistic is described by the spread of its sampling distribution, which includes the sampling design and the size of the sample. Larger samples result in a smaller spread.

II. Sampling Distribution of a Sample Proportion

A. Assume a simple random sample of size n from a large population with proportion p with $\hat{p}$ equal to the proportion of the sample.

1. The mean of the sampling distribution of $\hat{p}$ is the same as p.

2. The standard deviation of the sampling distribution of ($\hat{p}$) is defined as: $\sigma_{\hat{p}} = \sqrt{\dfrac{p(1-p)}{n}}$.

3. The equation for standard deviation should only be used when the population is at least 10 times as large as the sample.

4. Since ($\hat{p}$) = p there will be no bias in the statistic.

5. Since n is in the denominator, the standard deviation will get smaller as the population increases.

6. A normal approximation to the sampling distribution of $\hat{p}$ can be used when $np \geq 10$ and $n(1 - p) \geq 10$.

Example

A simple random sample of 2,300 high school students was asked whether they planned to apply for admission to an in-state college. In truth, 23% of all high school students applied to an in-state college (2 million high school students countrywide). What is the probability that the random sample of 2,300 students will yield a result within 3 percentage points of its true value?

➤ Is the population at least 10 times as large as the sample? Yes, a population of 2 million is more than 10 times the sample size.

➤ The standard deviation is

$$\sigma_{\hat{p}} = \sqrt{\frac{p(1-p)}{n}} = \sqrt{\frac{0.23(0.77)}{2300}} = 0.0088$$

➤ Can the normal approximation be used? Yes.

- Is $np \geq 10$ and $n(1 - p) \geq 10$?

$np = 2300(0.23) = 529$

$n(1 - p) = 1771$

➤ The area needed to be located under the normal distribution is $0.20 \leq \hat{y} \leq 0.26$. The z-score formula to standardize the interval is:

$$z = \frac{x - \text{mean}}{s}$$

$$z = \frac{0.20 - 0.23}{0.0088} = -3.4$$

$$z = \frac{0.26 - 0.23}{0.0088} = 3.4$$

$P(0.20 \leq \hat{y} \leq 0.26) = P(-3.4 \leq Z \leq 3.4) = 0.9997 - .0003 = 0.9994$ (use normal table)

There is more than a 99% probability that the results will be within 3 percentage points.

III. Sampling Distribution of a Sample Mean

A. Means are used for quantitative values rather than for looking at a proportion of a particular population. For example, knowing the mean national income, or the mean number of car accidents per day, makes the mean a key statistical inference.

B. If $\bar{x}$ is the mean of a simple random sample of size n from a large population with mean μ and standard deviation σ, the mean and standard deviation can be defined as the following:

$$\text{Mean} = \mu_{\bar{x}} - \mu$$

$$\text{Std Dev} = \sigma_{\bar{x}} = \frac{\sigma}{\sqrt{n}}$$

1. Since $\mu_{\bar{x}} - \mu$ there will be no bias in the statistic.

2. Since *n* is in the denominator, the standard deviation will get smaller as the population increases.

3. A normal approximation applies to the sampling distribution of $\mu_{\bar{x}}$.

Example

In the developed world, the average weight of a newborn baby is 7.5 pounds with a standard deviation of approximately 2.0. Suppose that in a hospital 25 babies are born over a week's period. What will be the mean and standard deviation of the weights of these babies? What is the probability that a baby will weigh greater than 10 pounds when born at this particular hospital?

➤ The mean will be 7.5 since $\mu_{\bar{x}} - \mu$.

➤ The standard deviation will be $\sigma_{\bar{x}} = \dfrac{\sigma}{\sqrt{n}} = \dfrac{7.5}{\sqrt{25}} = 1.5$

➤ Using the standard normal table to find the possibility of a baby weighing more than 10 pounds, where $P(X > 10)$.

$$z = \frac{x - \text{mean}}{s}$$

$$z = \frac{10 - 7.5}{1.5} = 1.67$$

Using the standard normal table

$$1 - .9525 = 0.0475$$

There is a 4.75% chance of a baby weighing greater than 10 pounds.

IV. **The Central Limit Theory**

A. In a simple random sample of size *n*, from any population with mean μ, and standard deviation σ, when *n* is large, the distribution of sample means is close to a normal distribution with mean μ and standard deviation $\dfrac{\sigma}{\sqrt{n}}$.

 V. **Sampling Distribution of a Difference Between Two Independent Sample Proportions**

A. At times, sampling distributions will involve the comparison of two populations.

B. The formula for the standard deviation of the difference between two sample proportions is equal to the following:

$$\sigma_{p_1-p_2} = \sqrt{\frac{p_1(1-p_1)}{n_1} + \frac{p_2(1-p_2)}{n_2}}$$

p_1 = sample 1, n_1 = sample size 1

p_2 = sample 2, n_2 = sample size 2

1. The set of differences of sample proportions is approximately normally distributed.

Example

In a simple random sample of people living in the northern New Jersey town of Ridgewood, 62% of 100 baseball fans are New York Yankees fans, while 51% of fans (90 samples) in the central New Jersey town of Princeton are Philadelphia Phillies fans. What is the probability that the difference in the percentage of Yankees fans to Phillies fans is more than 13%?

$$\sigma_{p_1-p_2} = \sqrt{\frac{p_1(1-p_1)}{n_1} + \frac{p_2(1-p_2)}{n_2}}$$

$$\sigma_{p_1-p_2} = \sqrt{\frac{0.62(0.38)}{100} + \frac{0.51(0.49)}{90}} = 0.07$$

$$z = \frac{x - mean}{s}$$

$$z = \frac{0.13 - 0.11}{0.07} = 0.29$$

Using the standard normal table

$$1 - .6141 = 0.3859$$

There is a 38.59% chance of the difference being more than 13%.

VI. Sampling Distribution of a Difference Between Two Independent Sample Means

A. At times, sampling distributions will involve the comparison of two means.

B. The set of differences in the sample means is approximately normally distributed.

C. The mean of the set of differences of sample means equals $\mu_1 - \mu_2$, which is the difference in population means.

D. The formula for the standard deviation of the difference between two sample means is equal to the following:

$$\sigma_{\bar{x}_1 - \bar{x}_2} = \sqrt{\frac{\sigma_1^2}{n_1} + \frac{\sigma_2^2}{n_2}}$$

Example

The distribution of men's heights is normal, with a mean of 69.0 inches and a standard deviation of 2 inches. The distribution of women's heights is normal, with a mean of 63.6 and a standard deviation of 2.5 inches. For a sample of 36 men and 36 women, what is the probability that the difference in mean heights between the men and women is greater than 6 inches?

Solution:

$$z = \frac{(\bar{x}_1 - \bar{x}_2) - (\mu_1 - \mu_2)}{\sqrt{\dfrac{\sigma_1^2}{n_1} + \dfrac{\sigma_2^2}{n_2}}}$$

$$= \frac{6 - 5.4}{\sqrt{\dfrac{2^2}{36} + \dfrac{2.5^2}{36}}} = \frac{0.6}{\sqrt{\dfrac{10.25}{36}}} \approx 1.12$$

Using the standard normal distribution table, 1 − 0.8686 = 0.1314. Thus, there is a 13.14% chance that the difference in mean heights between the men and women will exceed 6 inches.

Many of the calculations involving sampling distribution are complicated. You must have a good grasp of the Normal distribution and how to use z-scores in order to get these questions correct.

PART VI:

STATISTICAL INFERENCE

Estimation

I. Logic of Confidence Intervals, Meaning of Confidence Levels and Confidence Intervals, and Properties of Confidence Intervals

A. A confidence interval (C) can be calculated from data in the form of estimate ± margin of error.

B. A confidence level gives the probability that the interval will be within the true parameter value in repeated samples. It is a measure of the rate of success.

C. Typical confidence levels are 90%, 95%, or 99%.

Example

A math teacher proposes to use his AP® Calculus BC exam results to prove that he is a good instructor. He has taught AP® Calculus BC for the past 25 years and has instructed over 1,000 students. His AP® Statistics colleague takes a simple random sample of 50 students, AP® scores and finds a mean of 4.2 and a standard deviation of 0.3 (the highest AP® exam score possible is a 5). What can the teacher say about the mean score for all 1,000 students in the population?

➤ The mean of the sample is the same as the mean of population, and the standard deviation can be found using the formula below.

$$\mu_{\bar{x}} = \mu = 4.2$$

$$\sigma_x = \frac{\sigma}{\sqrt{n}} = \frac{0.3}{\sqrt{50}} = 0.04$$

➤ Since the distribution is close to normal based on the central limit theorem, the 68–95–99.7 rule states the following:

- About 95% of all samples will fall within two standard deviations of the mean.

    $4.2 + 0.08 = 4.28$

    $4.2 - 0.08 = 4.12$

    What we can say is that approximately 95% of all students' scores fall within the range of 4.12–4.28.

    4.2 ± 0.08 is the 95% confidence interval (C = 0.95) for μ. 0.08 can be considered the margin of error.

II. Confidence Interval for a Mean (Unknown), if the Standard Deviation Is Known

A. The calculation of a confidence interval for an unknown mean requires the following:

1. SRS or simple random sample

2. Normality

3. Independence

B. To find the confidence intervals, z-scores must be used. For example, if you wanted to construct the confidence interval of 90%, 95%, and 99%, you would need to calculate the upper and lower tail z-score limits via the formula below.

$$\text{Probability} = \frac{(1-C)}{2}$$

C. Critical value of the distribution are the *z*-scores that mark off a specified area under the standard normal curve.

Confidence Interval	Tail Area	Left *z*-score	Right *z*-score
90%	Probability $= \dfrac{(1-C)}{2} = \dfrac{1-0.9}{2} = 0.05$	−1.645	1.645
95%	Probability $= \dfrac{(1-C)}{2} = \dfrac{1-0.95}{2} = 0.025$	−1.960	1.960
99%	Probability $= \dfrac{(1-C)}{2} = \dfrac{1-0.99}{2} = 0.005$	−2.575	2.575

D. The formula for determining the confidence interval from a population having an unknown mean and a known standard deviation is:

$$\bar{x} \pm z \frac{\sigma}{\sqrt{n}}$$

Example

A size 5 soccer ball must have a circumference of 68-70 cm in order to be officially used by FIFA (Fédération Internationale de Football Association). The mean circumference of 100 random balls from a normally distributed population of 10,000 was 68.6 with a standard deviation of 15.4. Construct and interpret a 90% confidence interval for the mean circumference of all 10,000 balls in the population.

$$\bar{x} \pm z \frac{\sigma}{\sqrt{n}}$$

$$68.6 \pm 1.645 \frac{15.4}{\sqrt{100}}$$

$$68.6 \pm 2.53 = (66.07, \ 71.13)$$

There is a 90% confidence that the balls produced (10,000) will have a circumference from 66.07 to 71.13.

III. Confidence Interval for a Mean (Unknown), if the Standard Deviation Is also Unknown

A. The calculation of a confidence interval for an unknown mean and an unknown standard deviation requires the following:

1. SRS or simple random sample

2. Normality

3. Independence

B. Standard Error—if the standard deviation is estimated, the resulting statistic is called standard error. The standard error is calculated by the equation:

$$\frac{s}{\sqrt{n}}$$

s = sample standard deviation

The resulting statistics represent a t-distribution because it is *not* normal. The t-distribution is different for each sample size and specified by giving it degrees of freedom or *df*. *df* is calculated by $n - 1$. A t-distribution is similar to a normal distribution. t-distributions have the following characteristics:

1. They are symmetric about zero, unimodal, and bell-shaped.

2. The spread of a t-distribution is slightly larger.

3. As the degrees of freedom increase, the curve looks more normal. As the sample size increases, the sample standard deviation estimates move more closely to σ.

C. The formula for determining the confidence interval from a population having an unknown mean and an unknown standard deviation is:

$$\bar{x} \pm t \frac{s}{\sqrt{n}}$$

t = critical t-values for $t(n - 1)$ distribution (must be looked up in t-critical values table)

D. *t*-Distribution

Example

Below is a graph of the *t*-distribution for different degrees of freedom, with a comparison to the normal distribution:

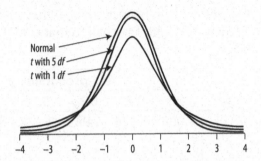

1. What value of the *t*-score cuts off the highest 10% of a *t*-distribution with 9 degrees of freedom?

   Answer: Cross-reference the row that contains 9 degrees of freedom (*df*) with the column that contains 0.10, as shown below.

df	Tail probability *p*					
	0.25	0.20	0.15	0.10	0.05	...
1	1.000	1.376	1.963	3.078	6.314	...
.	.	.	.	.	.	...
.	.	.	.	.	.	...
.	.	.	.	.	.	...
7	0.711	0.896	1.119	1.415	1.895	...
8	0.706	0.889	1.108	1.397	1.860	...
9	0.703	0.883	1.100	1.383	1.833	...
10	0.700	0.879	1.093	1.372	1.812	...
11	0.697	0.876	1.088	1.363	1.796	...
.	.	.	.	.	.	...
.	.	.	.	.	.	...
.	.	.	.	.	.	...

➤ The corresponding value of *t* is 1.383. An equivalent interpretation is that 1.383 represents the 90th percentile of a *t*-distribution with 9 degrees of freedom.

2. What value of the *t*-score cuts off the lowest 5% of a *t*-distribution with 10 degrees of freedom?

Answer: Cross-reference the row that contains 10 degrees of freedom (*df*) with the column that contains 0.05, as shown below.

df	Tail probability *p*					
	0.25	0.20	0.15	0.10	0.05	...
1	1.000	1.376	1.963	3.078	6.314	...
.	.	.	.	.	.	...
.	.	.	.	.	.	...
.	.	.	.	.	.	...
7	0.711	0.896	1.119	1.415	1.895	...
8	0.706	0.889	1.108	1.397	1.860	...
9	0.703	0.883	1.100	1.383	1.833	...
10	0.700	0.879	1.093	1.372	1.812	...
11	0.697	0.876	1.088	1.363	1.796	...
.	.	.	.	.	.	...
.	.	.	.	.	.	...
.	.	.	.	.	.	...

➤ Since the area under consideration is the lowest 5%, the correct value of *t* is –1.812.

➤ This means that the *t* value of –1.812 represents the 5th percentile of a *t*-distribution with 10 degrees of freedom.

IV. Confidence Interval for a Mean—Small Sample

Example

In a medium-sized city in Michigan, a survey of 16 schools was conducted to determine the number of teachers in each school statewide. The results showed a mean of 35 with a sample standard deviation of 5. Assuming that the population of the number of teachers in all Michigan schools is normally distributed, what is the 95% confidence interval for the mean number of teachers in all Michigan schools?

Answer: Since the sample size is less than 30, the formula to use is $\bar{x} \pm (t_c)\left(\dfrac{s}{\sqrt{n}}\right)$. The value of t_c can be found by using the t-distribution table value that corresponds to 15 degrees of freedom and a two-tailed 95% confidence interval. The value of t_c is 2.131.

➤ Thus, the 95% confidence interval is $35 \pm (2.131)\left(\dfrac{5}{\sqrt{16}}\right) \approx$ 35 ± 2.664.

In interval form, the answer is (32.336, 37.664).

V. Paired t Procedures

A. When comparing the responses of two treatments on the same subject, apply one-sample t procedures to the observed differences. The formula below is based on differences between the two treatments:

$$\bar{x}_{diff} \pm t \frac{s_{diff}}{\sqrt{n}}$$

VI. Conditions for Inference when Dealing with a Proportion

A. The calculation of a confidence interval for proportion requires the following:

1. SRS or simple random sample

2. Normality

3. Independence

B. The approximate level C confidence interval for a large population with an unknown proportion of p success is:

$$\hat{p} \pm z\sqrt{\frac{\hat{p}(1-\hat{p})}{n}}$$

VII. Confidence Interval for Difference for Two Population Proportions

The formula for computing the confidence interval for the difference between two population proportions p_1 and p_2 is given by the following: $(\hat{p}_1 - \hat{p}_2) \pm z_c\sqrt{\frac{(\hat{p}_1)(1-\hat{p}_1)}{n_1} + \frac{(\hat{p}_2)(1-\hat{p}_2)}{n_2}}$,

where z_c is the critical z-value of the normal distribution, $\hat{p}_1$ and $\hat{p}_2$ are the respective sample proportions, and n_1 and n_2 are the respective sample sizes. There are a few conditions that must be met in order to use this formula:

(a) The samples must be independent, so that observations in one sample affect the observations in the other sample.

(b) $\hat{p}_1 > \hat{p}_2$. If this is not the case, switch the labeling of sample 1 and sample 2.

(c) The sample sizes must be large enough so that each of $n_1\hat{p}_1$, $n_1(1 - \hat{p}_1)$, n_2p_2, and $n_2(1 - \hat{p}_2)$ must be at least 10.

(d) The population sizes must be at least 10 times as large as their respective sample sizes.

Example

A researcher is interested in determining the difference between the proportion of high school seniors in rural districts who are enrolled in AP® Statistics classes versus the proportion of high school seniors in urban districts who are enrolled in AP® Statistics classes. The samples are to be drawn from a populous state. In a sample of 250 rural high school seniors, 82 are enrolled in AP® Statistics. In a sample of 280 urban high school seniors, 56 are enrolled in AP® Statistics. What is the 90% confidence interval for the actual difference between the proportion of high school seniors enrolled in AP® Statistics in rural and urban areas.

Answer: All the necessary conditions are satisfied. We determine that $\hat{p}_1 = \dfrac{82}{250} = 0.328$ and $\hat{p}_2 = \dfrac{56}{280} = 0.2$. The critical z-value for a 90% interval is 1.645.

➤ Thus, the 90% confidence interval is

$$(0.328 - 0.2) \pm 1.645 \sqrt{\frac{(0.328)(1-0.328)}{250} + \frac{(0.2)(1-0.2)}{280}} .$$

This expression simplifies to approximately

$$0.128 \pm 1.645\sqrt{0.00088 + 0.00057} \approx 0.128 \pm 0.0627.$$

We can also write this answer in interval form as (0.0653, 0.1907).

➤ Based on these sample results, we can conclude that we are 90% confident that the actual difference between the proportion of rural high school seniors taking AP® Statistics is between 6.5% and 19.1% higher than the proportion of urban high school seniors taking AP® Statistics.

VIII. Confidence Interval for Slope of a Least Squares Regression Line

A. A level C confidence interval with standard error (SE) for the slope of a regression line is defined as:

$$b \pm tSE_b, \text{ where } SE_b = \frac{s}{\sqrt{\sum(x - \bar{x})^2}}$$

Less math and more understanding will assist you in this portion of the AP® test. Many questions require you to explain what the confidence interval means with regard to the distribution being studied. Calculations regarding confidence intervals will not be difficult to complete. Memorization of the equations is not required since they can be found on the AP® Statistics reference sheet. The standard formula for any confidence interval is:

Confidence Interval = statistics ±
(critical value) • (standard deviation of statistic)

Single-Sample

Statistic	Standard Deviation of Statistic
Sample Mean	$\dfrac{\sigma}{\sqrt{n}}$
Sample Proportion	$\sqrt{\dfrac{p(1-p)}{n}}$

Two-Sample

Statistic	Standard Deviation of Statistic
Difference of sample means	$\sqrt{\dfrac{\sigma_1^2}{n_1} + \dfrac{\sigma_2^2}{n_2}}$ When $\sigma_1 = \sigma_2$ $\sigma\sqrt{\dfrac{1}{n_1} + \dfrac{1}{n_2}}$
Difference of sample proportions	$\sqrt{\dfrac{p_1(1-p_1)}{n_1} + \dfrac{p_2(1-p_2)}{n_2}}$ When $p_1 = p_2$ $\sqrt{p(1-p)}\sqrt{\dfrac{1}{n_1} + \dfrac{1}{n_2}}$

Tests of Significance

I. Test of Significance

A test of significance is a common type of statistical inference that will assess evidence provided by data about some claim concerning a population.

A. The basis behind a significance test is that an outcome that would rarely occur if a particular claim were true, provides good support that the claim is *not* true.

II. Statistical Tests

A good statistical test always starts with a careful statement of claims using *Null and Alternative Hypotheses*. The null hypothesis and alternative hypothesis refer to populations and not outcomes.

A. Null Hypothesis or H_0—the statement that is being tested. A significance test is used to measure the validity of the null hypothesis, by testing the evidence *against* it. The null hypothesis is a statement of "no difference" or "no effect."

B. Alternative Hypothesis or H_a is the claim against the population at which the statistical test is aimed.

Example

A certain SAT prep course advertises that students will increase their mathematics score by an average of at least 50 points if students attend a monthly class.

H_0: students who take this particular SAT course will not show at least a 50-point increase in their mathematics SAT score after 1 month. $\mu < 50$

H_a: students who take this particular SAT course will show at least a 50-point increase in their mathematics SAT score after 1 month. $\mu > 50$

III. P-values

P-values refer to a probability, based on H_0 being true, that an outcome that is observed would take a value as extreme as, or more extreme than, that actually observed. The smaller the P-value mathematically, the stronger the evidence against H_0.

A. One-sided P-value

Example

Assume that an SAT prep course company has had over 500 students attend classes nationwide. The president of the company wants to test its claim by sampling 100 students. When the data is amassed it is found that the average mathematics SAT score increased 60 points with a $\sigma = 35$.

H_0: students who take this particular SAT course will not show at least a 50-point increase in their mathematics SAT score after 1 month. $\mu < 50$

H_0: students who take this particular SAT course will show at least a 50-point increase in their mathematics SAT score after 1 month. $\mu > 50$

$$\text{Test Statistic or } z = \frac{\text{estimate} - \text{hypothesized value}}{\text{standard deviation of the estimate}}$$

$$\text{Test Statistic or } z = \frac{60 - 50}{\frac{35}{\sqrt{100}}} = 2.86$$

The *P*-value is the probability of getting at least one value as extreme as possible if H_0 were true. In this case, $P(\bar{x} \geq 60)$. The *z*-score for 2.86 is equivalent to a 99.7% chance that the president would obtain a sample of 100 students who gained at least 50 points after taking the course (H_0 is rejected).

You will be tested on your ability to correctly write or recognize the null and alternate hypotheses. You need to be able to use the z-score formula for any statistical inference question on the free-response section.

IV. Type I Errors

A. H_0 is rejected, but H_0 is actually *true*.

B. Alpha (α) risk or probability of committing a Type I error

V. Type II Errors

A. H_0 is not rejected when H_0 is *false*.

B. Beta (β) risk or probability of committing a Type II error

C. Type II errors occur because the statistical test used is not sensitive enough to detect the alternative.

D. The *Power* of the test against the alternative is 1 minus the probability of a Type II error or Power = $1 - \beta$.

VI. Statistical Significance

A. Compares *P* with a fixed value that is known. The decisive value of *P* is known as the significance level or alpha (α). If $\alpha = 0.05$, the evidence requires that data against H_0 be strong enough that it would occur no more than 5% of the time. If $\alpha = 0.01$, then stronger evidence against H_0 is required.

VII. Sample Test for a Proportion

A. The calculation of a large sample test requires

1. a SRS or simple random sample

2. a sample size that is less than 10% of the population

3. independence

Example

A certain commercial on television states that 9 out of 10 people respond positively to a certain dietary shake for losing weight. You think the proportion is lower and you run a hypothesis test at 10% significance level. What is the conclusion if 82 out of 90 of an SRS sample say they responded positively?

$$H_0: p \geq 0.9$$

$$\alpha = 0.10$$

$$H_a: p < 0.9$$

$$s_p = \sqrt{\frac{p(1-p)}{n}}$$

$$s_p = \sqrt{\frac{(.90)(.10)}{90}} = 0.032$$

The observed sample proportion is $\dfrac{82}{90} = 0.91$

$$\text{Test Statistic or } z = \frac{\text{estimate-hypothesized value}}{\text{standard deviation of the estimate}}$$

$$\text{Test Statistic or } z = \frac{0.90 - 0.91}{0.032} = -0.31 \text{ with a } p\text{-value of } 0.3783$$

Since the p-value > 0.10, we cannot reject H_0 at the 10% level of significance.

III. Test for a Difference Between Two Proportions

A. Requirements for a difference between two proportions:

1. SRS or simple random sample

2. Normality

3. Independence of samples

Example

In 2008, 500 out of 800 polled students at a particular school stated that they wanted more pizza available at lunch. One week later, a separate group of 1,000 students from the same school were polled and 730 responded that they wanted more pizza at lunch. At a 10% significance level, is there sufficient evidence that the popularity of pizza increased?

$H_0: p_1 - p_2 = 0$

$\hat{p}_1 = \dfrac{500}{800} = 0.625$

$H_a: p_2 - p_1 > 0$

$\hat{p}_2 = \dfrac{730}{1000} = 0.73$

$\hat{p} = \dfrac{500 + 730}{800 + 1000} = 0.68$

$$s_p = \sqrt{\frac{p(1-p)}{n}}$$

$$s_p = \sqrt{\frac{0.68(0.32)}{1800}} = 0.011$$

Test Statistic $z = \dfrac{0.73 - 0.625}{0.011} = 9.55$

Critical z value for $\alpha = 0.10$ is 1.28

> Since $9.55 > 1.28$ reject the null hypothesis. The popularity of pizza did increase.

IX. Hypothesis Test for a Mean (See chapter 15)

Formula $\quad \dfrac{\sigma}{\sqrt{n}}$

X. Hypothesis Test for Difference Between Two Means (See chapter 15)

Formula $\quad \sqrt{\dfrac{\sigma_1^2}{n_1} + \dfrac{\sigma_2^2}{n_2}}$

XI. Hypothesis Testing—Paired Sample Means

A manufacturer makes two different types of baseball bats, aluminum and ceramic. The company is interested in testing the hypothesis that the ceramic bats perform better than the aluminum ones at the 1% level of significance. Twenty-three Little League baseball players participated in this study. Each boy was given one bat of each type, with which he took 20 swings. The number of hits into the outfield was recorded for each bat, shown in the table the follows.

Aluminum	Ceramic	Difference (C–A)
4	5	1
2	5	3
3	3	0
3	6	3
6	8	2
3	6	3
5	6	1
7	12	5
5	8	3
9	15	6
6	8	2
3	2	–1
1	3	2
2	5	3
4	8	4
8	7	–1
8	8	0
3	6	3
2	5	3
2	4	2
6	11	5
5	9	4
4	6	2

The mean of the paired differences (μ_d) is calculated as approximately 2.3913, and the corresponding standard deviation (σ_d) as approximately 1.8275. The null hypothesis H_0 states that $\mu_d \leq 0$, whereas the alternative hypothesis H_a states that $\mu_d > 0$. The test statistic $t = \dfrac{2.3913 - 0}{1.8275 / \sqrt{23}} \approx 6.275$.

Using the t-distribution table, with 22 degrees of freedom, the corresponding t value for the one tail 1% level of significance is 2.508. Since 6.275 > 2.508, we must reject H_0. Our conclusion is that the ceramic bats will produce a higher average number of hits into the outfield than the aluminum ones.

XII. Chi-Square

A. Chi-square is a goodness-of-fit test to determine whether a population has a certain hypothesized distribution. The individuals in the population are expressed as proportions of individuals falling into various outcome categories. Since chi-square is a statistical test of significance, all rules of a significance test must be followed.

B. Formula/Set-up for Chi-Square

H_0: the actual population proportions are equal to the hypothesized proportions

H_a: the actual population proportions differ from the hypothesized proportions

$$X^2 = \sum \frac{(\text{observed count} - \text{expected count})^2}{\text{expected count}} = \sum \frac{(O - E)^2}{E}$$

X^2 has an approximate distribution of k-1 degrees of freedom. The test is used with critical values from the chi-square distribution table, based on the degrees of freedom.

C. Chi-Square analysis one-way table

A biologist is conducting a mating experiment between two fruit flies. One fruit fly has a normal wing structure, while the other has a slightly deformed wing structure called "dumpy."

Null Hypothesis—there is no statistical difference between expected data and observed data

Alternative Hypothesis—there is a statistical difference between expected data and observed data

$$X^2 = \sum \frac{(\text{observed count} - \text{expected count})^2}{\text{expected count}} = \sum \frac{(O-E)^2}{E}$$

o = observed number of individuals

e = expected number of individuals

Σ = sum of values

Degrees of freedom = expected phenotypes –1

D. Use of the Chi-Square Table of Critical Values

Probability (p)	Degrees of Freedom				
	1	2	3	4	5
0.05	3.84	5.99	7.82	9.49	11.1

If the calculated chi-square is greater than or equal to the critical value, the null hypothesis is rejected with a reassurance of 95%, and you would see the null hypothesis as being correct only 5% of the time.

Sample Data

Phenotype	# Observed	# Expected	(o–e)	(o–e)²	$\frac{(o-e)^2}{e}$
Normal wing	70	75	–5	25	0.33
Dumpy wing	30	25	5	25	1.00
					1.33

Result: There is no difference between observed and expected phenotypes. Thus, we cannot reject the null hypothesis.

XIII. **Chi-Square Two-Way Tables**

A. Two-way tables allow for multiple comparisons to be made, at one time, with some overall measure of confidence.

B. Multiple comparisons using chi-square require:

1. An overall test to see if there is any difference between different parameters, and

2. A follow-up analysis to measure if the parameters are different.

C. A two-way table will show both successes and failures with *r*-rows and *c*-columns, giving relationships between two categorical variables.

Example

Suppose a vendor at a baseball game wants to make an inference about the items he sells at specific times in the game. The table below measures the amount of items sold during 3-inning intervals of a baseball game.

Item Sold	Innings 1–3	Innings 4–6	Innings 7–9	Total
Hot Dogs	45	32	25	102
Sodas	34	56	10	100
Popcorn	32	32	70	134
TOTAL	111	120	105	336

H_0: There is no difference in the distribution of items sold based on the intervals of 3 innings each.

H_a: There is a difference in the distribution of items sold based on the intervals of 3 innings each.

The expected counts can be calculated using the Expected Cell Count formula:

$$\text{Expected Count} = \frac{\text{row total} \times \text{column total}}{n}$$

To find the expected cell count for hot dogs sold in innings 1–3, use the data in the previous table.

$$\text{Expected Count} = \frac{\text{row total} \times \text{column total}}{n}$$

$$\text{Expected Count} = \frac{102 \times 111}{336} = 33.7$$

Item Sold	Innings 1–3	Innings 4–6	Innings 7–9	Total
Hot Dogs	33.7	36.4	31.8	101.9
Sodas	33.0	35.7	31.3	100
Popcorn	44.3	47.9	41.9	134.1
TOTAL	111	120	105	336

Using the chi-square formula for independence, the statistic is calculated for each cell and summed over the nine cells.

$$X^2 = \sum \frac{(\text{observed count} - \text{expected count})^2}{\text{expected count}} = \sum \frac{(O - E)^2}{E}$$

$$X^2 = \frac{(45 - 33.7)^2}{33.7} + \frac{(32 - 36.4)^2}{36.4} + \frac{(25 - 31.8)^2}{31.8}$$

$$+ \frac{(34 - 33)^2}{33} + \frac{(56 - 35.7)^2}{35.7} + \frac{(10 - 31.3)^2}{31.3}$$

$$+ \frac{(32 - 44.3)^2}{44.3} + \frac{(32 - 47.9)^2}{47.9} + \frac{(70 - 41.9)^2}{41.9}$$

$$= 59.38$$

Because there are 3 rows and 3 columns, the degrees of freedom for each is 2. Since sales and innings are independent, multiply the degrees of freedom to get 4.

The chi-square critical value is 9.49 at 0.05 p-valve. Reject the null hypothesis since 59.38 > 9.49. Furthermore, 59.38 does not fall below any p value in the chi-square distribution of critical values.

 Chi-square analysis has been on every released AP® Statistics test. Make sure you can use the chi-square critical value table on the test correctly.

XIV. Hypothesis Test for Slope of Least Squares Lines

A. Use null hypothesis and alternate hypothesis.

1. H_0: There is no linear relationship between 2 variables or $b = 0$.

2. H_a: There is a linear relationship between 2 variables or b does not $= 0$.

B. Assumptions for inference for the slope of the least squares equation:

1. Sample is randomly selected.

2. The scatterplot should be linear.

3. No pattern in the residual plot (thus linear).

4. Normal distribution of the residuals.

Formula to be used is the t-test of a slope

$$t = \frac{b}{SE_b}$$

b = slope

SE_b = standard error of the slope

Hypothesis Test—Slope of Least Squares Regression Line

The hypothesis test for the slope of a least squares regression line allows us to determine if there is a statistically significant relationship between the independent (x) and dependent (y) variables. If there is a linear relationship between x and y, then the slope should not equal zero.

Example

The heights (in inches), x, and shoe sizes, y, of 15 women are shown in the following table.

x	63	62	62	63	68	65	68	69	70	62	60	60	61	63	64
y	7.5	7	7.5	6.5	8	7.5	9	9	10	6	5	5.5	6	7	7.5

The least squares regression equation is $y = -0.390x - 17.7$. Is the slope of the least squares regression line different from zero at the 1% level of significance?

Answer: Let b represent the slope of the regression line. Then we have

H_0: $b = 0$, which means that there is no significant linear relationship between x and y.

H_a: $b \neq 0$, which means that there is a relationship between the two variables.

To find the critical t value, we use $t = \dfrac{b}{SE_b}$. We know that $b = 0.390$. The calculation for SE_b is found by $\dfrac{\sqrt{\dfrac{\sum(y - \hat{y})^2}{n-2}}}{\sqrt{(x - \bar{x})^2}}$,

where y represents the actual value of the dependent variable, $\hat{y}$ represents the predicted value of y, based on the regression equation, $\bar{x}$ represents the mean of the x values (64), and n is the number of paired data (15). For example, when $x = 70$, $y = 10$, and, $\hat{y} = 0.390(70) - 17.7 = 9.6$. By substitution, we find that $\dfrac{\sqrt{\dfrac{\sum (y - \hat{y})^2}{n-2}}}{\sqrt{(x - \bar{x})^2}} = \dfrac{\sqrt{\dfrac{3.619}{13}}}{\sqrt{150}} \approx 0.043$.

Then the critical $t = \dfrac{0.390}{0.043} \approx 9.07$. Using the t-distribution table for $n - 2 = 13$ degrees of freedom, we find that $t = 3.012$ in the column for a two-tailed 1% level of significance. Since $9.07 > 3.012$, we must reject H_0. Our conclusion is that there is a linear relationship between the heights of women and their shoe sizes.

Test Tip

The hypothesis test for slope of least squares lines will require you to possibly comment on r or r². r is the correlation coefficient while r² is the coefficient of determination (which is the comparison of the least squares regression line to the data). If r = 0.70, then r² = 0.49, which indicates that 49% of the variance in one variable is accounted for by the other variable.

PART VII:

STRATEGIES AND PRACTICE FOR THE EXAM

Hints for the Exam

I. Tips for the Multiple Choice-Questions

➤ Since you can use a calculator throughout both sections of the AP® exam, make sure to bring the calculator you used all year. Familiarity with a key piece of technology will benefit you.

➤ Many of the multiple-choice questions are multi-level. Multi-level means that the questions have several parts and could pull concepts from multiple parts of the AP® outline. For example, there may be a question based on Sampling/ Experimentation and Exploring Data such as measurements of center.

➤ *Do not* memorize any formulas for the test. Instead, download the AP® Statistics Course and Exam Description at *www.collegeboard.org/ap* to access the exam's formula sheet. If you are unsure about the equations, match them up to the various chapters in this *Crash Course*.

➤ There are several key words to look for on multiple-choice questions (you should also assimilate these words when answering free-response questions). The key words include those in the following grid:

Key Words Used in AP® Statistics Multiple-Choice Items

1st quartile	experiment	percentile
3rd quartile	frequency histogram	pie chart
alternate hypothesis	geometric distribution	placebo
bar graph	interquartile range	population
binomial distribution	least squares estimation	proportion
blocking	level of significance	qualitative
boxplot	linear regression	quantitative variable
categorical variable	margin of error	range
census	mean	sampling distribution
chi-square	measure of center	simple random sample
confidence interval	measure of shape	standard deviation
confidence level	measure of spread	standard error
control	median	stem-leaf plot
correlation	mode	stratified random sample
critical value	mutually exclusive	*t*-test
distribution	normal distribution	transformation
dotplot	null hypothesis	variance

II. Tips for the Free-Response Questions

Typically, the free-response questions are relatively predictable. Questions 1–5 increase in difficulty as you read along, and question 6 is the Investigative Task. Question 6 is worth almost twice as much as the other individual questions, so do not skip it.

FREQUENT CONCEPTS FOUND ON QUESTIONS 1–5

➤ Plotting and interpreting data from a plot—discussing shape, center, and spread

➤ Defining characteristics of good experiments and sampling techniques

➤ Deciding between which is a better experiment or sampling method when comparing various examples

➤ A question relating to the normal distribution

➤ Using a statistical inference test to indicate if there is a difference in means

QUESTION 6—THE INVESTIGATIVE TASK

➤ You will have to demonstrate your understanding of a variety of topics, and your ability to integrate statistical ideas and apply them in a new context, or in an unusual way. For example, you may have to link the concept of data transformation to sampling techniques. You may have to analyze data from a computer printout and make conclusions. Question 6 only seems hard because of the multiple steps involved—this *Crash Course* book equips you to handle it.

III. Facts and Concepts

Study the facts and concepts below since they make up the majority of the test. Review your in-class problems and correlate them to the *Crash Course* to maximize your understanding.

EXPLORING DATA

➤ When commenting on a distribution, be sure to include shape, center, and spread.

➤ Comment on outliers.

➤ Mean and median are measures of center.

- If the distribution is skewed to the right, then the mean is greater than the median.

- If the distribution is skewed to the left, then the mean is less than the median.

- Do *not* say a distribution is *normal* because it is symmetrical and unimodal.

- Use the word *normal* only when appropriate.

- Adding a positive number to a data set will increase the mean and median. The standard deviation and interquartile range *do not change.*

- Subtracting a positive number from a data set will decrease the mean and median. The standard deviation and interquartile range *do not change.*

- Multiplying all numbers in the data set by a constant k multiplies the mean, median, interquartile range, and standard deviation by the constant.

➤ When plotting your data and looking for correlations, remember the following:

- Correlations measure both direction and strength of a linear relationship.

- Strong positive correlations have a positive slope and an r value close to 1.

- Strong negative correlations have a negative slope and an r value close to -1.

- No correlation means $r = 0$.

- Residual plots help us to know how well the regression line fits the data.

SAMPLING AND EXPERIMENTATION

➤ Simple Random Samples

- Each individual has an equal chance of being selected.

- Each individual and the probability of being chosen is independent of others in the population.

➤ Well-designed experiments have the following:

- Controls

- Randomization

- Replication

➤ Lurking Variables

- Something that has an effect on your response, and should have been included, but was not represented in the analysis.

- For example, suppose you were studying the effects of the following on colon cancer—diet, age, smoking habits, and occupation—but you omitted gender.

➤ Blocking

- Grouping of subjects based on a shared characteristic

- Blocking by gender, age, weight, or race

PROBABILITY

➤ Understand the following probability rules:

- When two events A and B have no outcomes in common (*disjoint*)—add their probabilities.

- If two events A and B are independent from one another (*independent*)—multiply their probabilities.

- The complement rule.

- Conditional Probability Rules

➤ Discrete Random Variables and Probability Distribution Tables

➤ Binomial and Geometric Probability Distributions

➤ The normal distribution

- Shape, and location of mean and median

- 68–95–99.7 Empirical Rule

- The use of the standard normal table

STATISTICAL INFERENCE

➤ You must be able to decide which statistical inference procedure is appropriate in a given setting.

➤ Know the difference between a population parameter, a sample statistic, and the sampling distribution of a statistic.

> Hypothesis Testing

- State hypotheses in words and symbols.
- Identify the correct inference procedure.
- Calculate the test statistic.
- Draw a conclusion in context that is directly linked to your test statistic.

> On any confidence interval problem:

- Identify the population of interest and the parameter.
- Choose the appropriate inference procedure and verify conditions for its use.
- Carry out the inference procedure.
- Interpret your results in the context of the problem.

> Type I error: Rejecting the null hypothesis when it is true

> Type II error: Failing to reject a null hypothesis when it is false

> Power of a test: Probability of correctly rejecting a null hypothesis

$$\text{Power} = 1 - P(\text{Type II error})$$

> Chi-Square test for goodness of fit

IV. Task Verbs Used in Free-Response Questions*

The following **task verbs** are commonly used in the free-response questions.

Calculate: Perform mathematical steps to arrive at a final answer: (e.g., algebraic expressions or diagrams with properly substituted numbers and correct phrasing). Example: Calculate the standard deviation of the random variable.

---

* Adapted from College Board AP® Statistics Course and Exam Description, 2019–2020, p. 239

Comment: Take a position on whether a statement is true with supporting evidence. Example: Comment on whether you believe that a company is short-changing customers.

Compare: Provide a description or explanation of similarities or differences. Example: Compare the distribution of grades in the classes of Mrs. Johnson and Mr. Crofton.

Construct/Complete: Represent data in graphical or numerical form. Construct a back-to-back stemplot comparing points scored by a football team in division and non-division games.

Describe: Provide the relevant characteristics of representations, distributions, or methods. Example: Describe the distribution of the life expectancy of a particular brand of microwave oven.

Determine: Apply an appropriate definition or perform calculations to identify values, intervals, or solutions. Example: Determine the probability that it will take more than 5 trials to draw a spade from a standard deck of cards.

Estimate: Use models or representations to find approximate values for functions. Example: Estimate the expected number of late flights in a particular week based on prior statistics.

Explain: Provide information about how or why a relationship, process, pattern position, situation, or outcome occurs, using evidence and/or reasoning to support or qualify a claim. Example: Explain why the sample may be biased.

Find a point estimate or interval estimate: Use models or representations to find approximate values for uncertain figures. Example: Based on the histogram, estimate the median of the data.

Give examples: Provide a specific example that meets given criteria. Example: Give examples of independent events.

Identify/Indicate/Circle: Indicate or provide information about a specific topic in works or by circling, shading, or marking given information without elaboration or explanation. Example: Shade the area under the normal curve that corresponds to the probability of a Type I error.

Interpret: Describe the connection between a mathematical expression, representation, or solution and its meaning within the realistic context of a problem, sometimes including consideration of units. Example: Interpret the value of r^2 within the context of the straight-line relationship between the weight of a car and its miles per gallon.

Justify: Provide evidence to support, quality, or defend a claim and/or provide statistical reasoning to explain how that evidence supports or qualifies the claim. Example: Does the evidence support the claim that the ocean temperature is warmer over the past 10 years? Justify your reasoning.

Verify: Confirm that the conditions of a particular definition, distribution, or inference method are met in order to verify that it is applicable in a given situation. Example: Verify that the conditions of the chi-square goodness of fit are met and then perform the test.

Practice Multiple-Choice Questions

Practice with the following AP®-style questions. Then go online to access our timed, full-length practice exam at *www.rea.com/ studycenter*.

1. A set of data is to be transformed using the following equation where an old score will be transformed into a new score.

    new score = 5 + 2.3(*old score*)

    Which of the following statements is INCORRECT?

    (A) New median = 5 + 2.3(old median)

    (B) New mean = 5 + 2.3(old mean)

    (C) New standard deviation = 5 + 2.3(old standard deviation)

    (D) New IQR = 2.3(old IQR)

    (E) New range = 2.3(old range)

2. An automobile parts manufacturer wants to inspect a sample of new brake pads before shipping them out. The brake pads are placed in boxes in the order they come off the assembly line. There are 100 boxes with each box containing 50 brake pads. The manager decides to number each box 00–99 and then randomly selects 3 boxes. Every brake pad in those 3 boxes will be inspected. What type of sampling method did he use?

    (A) Stratified random sample

    (B) Simple random sample

    (C) Convenience sample

    (D) Systematic sample

    (E) Cluster sample

3. The mean weight of an orange at a large orange orchard is 140 grams with a standard deviation of 10.6 grams. A random sample of 50 oranges is selected and weighed. Which of the following represents the probability that the sample mean will be more than 150 grams?

(A) $P\left(z > \dfrac{150-140}{\frac{10.6}{\sqrt{50}}}\right)$

(B) $P\left(z < \dfrac{150-140}{\frac{10.6}{\sqrt{50}}}\right)$

(C) $P\left(z > \dfrac{150-140}{10.6}\right)$

(D) $P\left(z > \dfrac{150-140}{\frac{10.6}{50}}\right)$

(E) $P\left(z > \dfrac{140-150}{\frac{10.6}{\sqrt{50}}}\right)$

4. Kelly is interested in buying a particular model used car. She wants to know roughly how much she should be paying based on the mileage of the car. She examines a sample of 10 cars and records the price in dollars and the mileage in miles for each car. A regression analysis table to predict the price of a used car from its mileage is given below.

	Coef.	SE Coef.	t Stat	P-value
Constant	24660.9136	2202.68082	11.1958634	3.6306E-06
Mileage	–0.2083181	0.06840524	–3.0453528	0.01593314

Which of the following is a correct interpretation of the slope in context?

(A) For every $1 that the price of a used car increases, the predicted mileage of a used car decreases by about 0.21 miles.

(B) For every 1 mile that the mileage on a used car increases, the predicted cost of the used car decreases by $0.21.

(C) For every 1 mile that the mileage on a used car increases, the predicted cost of the used car increases by $24,660.91.

(D) For every 0.21 miles that the mileage on a used car increases, the predicted price of the used car decreases by $1.

(E) For every $1 that the price of a used car increases, the predicted mileage of a used car increases by about 24,660.91 miles.

5. Police Officer Ron is conducting a study on the drug use of teenagers at the high school in the city where he works. The high school is made up of 4 classes each having freshmen, sophomores, juniors, and seniors—each having approximately the same number of students. He selects a random sample of 25 freshmen, 25 sophomores, 25 juniors, and 25 seniors to take part in the study. Officer Ron meets with each of the 100 students individually and asks them questions about their drug use. Which form of bias is present in his study?

(A) Non-response bias

(B) Under-coverage bias

(C) Voluntary response bias

(D) Response bias

(E) Sampling bias

6. An experiment is being conducted to see which of two razors gives a smoother, cleaner shave. The age of a man's skin is known to impact the smoothness of the shave. Thirty men of various ages are volunteering to take part in this experiment. Which of the following would be the best way to assign the two razors, A and B?

(A) Allow the 30 men to decide amongst themselves who gets Razor A and who gets Razor B.

(B) Give the 10 youngest men Razor A, the 10 oldest men Razor B, and split up the 10 men in the middle randomly—5 get Razor A and 5 get Razor B.

(C) Pair the men up, by taking the oldest man and the youngest man together for Pair 1, then the next oldest and next youngest for Pair 2, and so forth until 15 pairs are made. Then within each pair, one man will randomly get Razor A and one will get Razor B.

(D) Pair the men up. By taking the two youngest men as Pair 1, the next two youngest as Pair 2, and continuing until the two oldest men are paired together. Then, both men in Pair 1 will get Razor A, both men in Pair 2 will get Razor B, and so forth.

(E) Of the ten youngest men, half will randomly get Razor A and the other half Razor B. Of the ten oldest men, half randomly get Razor A and the other half Razor B. Finally, of the ten men in the middle, half get Razor A and the other half Razor B.

7. Measurements of soil quality were taken from a field that is near a nuclear power plant. Concentrations of a certain nuclear byproduct were obtained from 20 measurements taken at the surface of the field, 20 measurements were taken at 3 feet beneath the field, and 20 measurements were taken 6 feet beneath the field. What type of study was conducted and what is the response variable?

(A) An observational study was conducted and the response variable is the concentration of the nuclear byproduct.

(B) An observational study was conducted and the response variable is the depth of the soil.

(C) An experiment was conducted and the response variable is the concentration of the nuclear byproduct.

(D) An experiment was conducted and the response variable is the depth of the soil.

(E) A census was conducted and the response variable is the depth of the soil.

8. Suppose that in a certain part of the world, in any 10-year period, the probability that someone will get the flu virus is 0.12; the probability that someone will get a stomach virus is 0.25; and the probability that someone gets both the flu virus and a stomach virus is 0.08. What is the probability of getting a stomach virus, given that they have the flu virus?

(A) 0.03   (D) 0.25

(B) 0.67   (E) 0.29

(C) 0.32

9. The number of pairs of shoes that a vendor sells per day has the following probability distribution.

Shoes	0	1	2	3	4	5
Probability	0.12	0.20	0.35	0.17	0.10	0.06

If each pair of shoes costs $89, what is the expected daily total dollar amount taken in by the vendor from the sale of shoes?

(A) $2.11

(B) $118.37

(C) $187.79

(D) $161.09

(E) $1,335

10. The platypus is one of only a few mammals that lays eggs. When a female platypus lays two eggs, the survival rate of those babies is fairly low. The probability that the first egg hatches and survives is 0.26. If the first egg hatches and survives, the probability that the second egg also hatches and survives is 0.09. If the first egg doesn't hatch, then the probability that the second egg hatches and survives is 0.15. Which of the following probability distributions represents X, the number of surviving eggs when a female platypus lays two eggs?

(A)

X	1	2
Probability	0.6290	0.3476

(B)

X	1	2
Probability	0.3476	0.0234

(C)

X	0	1	2
Probability	0.6290	0.1110	0.0234

(D)

X	0	1	2
Probability	0.6290	0.2366	0.0234

(E)

X	0	1	2
Probability	0.6290	0.3476	0.0234

11. At a battery manufacturing plant, an inspector selects four batteries from each day's production and tests them to see if they are in working condition. If at least 3 of the 4 batteries are in working condition, he will conclude that the whole day's production is acceptable. If, in reality, only 92% of the batteries are in working condition, what is the probability the entire day's production is considered acceptable?

(A) 0.249

(D) 0.034

(B) 0.716

(E) 0.779

(C) 0.965

12. Suppose that a study revealed that the mean weight of a high school football player is 195 pounds and the standard deviation is 15.6 pounds. What is the standard deviation for the combined weight of 11 random high school football players?

(A) 51.74 pounds

(D) 2145 pounds

(B) 171.6 pounds

(E) 13.1 pounds

(C) 646.64 pounds

13. Suppose that 15% of cars on the road right now need an oil change. If 10 cars are chosen at random, which of the following is the probability that exactly 3 of them will need an oil change?

(A) $\binom{10}{3}(0.15)^3(0.85)^7$

(B) $(0.15)^3(0.85)^7$

(C) $\binom{10}{3}(0.15)^3$

(D) $\binom{10}{3}(0.15)^7(0.85)^3$

(E) $3(0.15)$

14. At a plant that produces paper cups, it is known that through the manufacturing process, 7% of cups do not meet design specifications. After making some changes to the manufacturing process, an inspector working at the plant wants to determine if less than 7% of paper cups produced do not meet design specifications. He plans to collect data on a random sample of 500 cups. What null and alternative hypotheses would he use to test his claim?

(A) $H_0 : p = 0.07$ *versus* $H_a : p > 0.07$

(B) $H_0 : p = 0.07$ *versus* $H_a : p \neq 0.07$

(C) $H_0 : p = 0.07$ *versus* $H_a : p < 0.07$

(D) $H_0 : \hat{p} = 0.07$ *versus* $H_a : \hat{p} < 0.07$

(E) $H_0 : p < 0.07$ *versus* $H_a : p = 0.07$

15. In a test on the hypotheses $H_0 : \mu = 14$; $H_a : \mu \neq 14$ the *p*-value was found to be 0.042. What conclusion should be made at the 5% level of significance?

(A) Since 0.042 < 0.05, the null hypothesis will be rejected. There is sufficient evidence to claim that the true mean is not 14.

(B) Since 0.042 < 0.05, the null hypothesis will be accepted. There is sufficient evidence to claim that the true mean is not 14.

(C) Since 0.042 < 0.05, the null hypothesis will be rejected. There is no sufficient evidence to claim that the true mean is not 14.

(D) Since 0.042 < 0.05, the null hypothesis will fail to be rejected. There is sufficient evidence to claim that the true mean is not 14.

(E) Since 0.042 < 0.05, the null hypothesis will fail to be rejected. There is no sufficient evidence to claim that the true mean is not 14.

16. The following set of hypotheses are to be tested: $H_0: \mu = 20$ *versus* $H_a: \mu > 20$. Which of the following scenarios would have the greatest amount of power?

    (A) In reality, $\mu = 21$ and $n = 50$.

    (B) In reality, $\mu = 21$ and $n = 500$.

    (C) In reality, $\mu = 26$ and $n = 50$.

    (D) In reality, $\mu = 26$ and $n = 500$.

    (E) In reality, $\mu = 18$ and $n = 500$.

17. In the state of Ohio, an official analyzed the goal differential between the winning team and losing team in every playoff game in 2018 for girls' soccer. The boxplot below represents the data he collected.

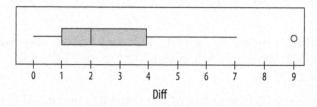

Diff

    The official wishes to discuss the center and spread of the data. Which of the following measures should he use to best represent this data?

    (A) Mean and standard deviation

    (B) Median and standard deviation

    (C) Mean and interquartile range

    (D) Median and interquartile range

    (E) Median and range

18. Maria wanted to investigate the mean amount of time a patient at her local hospital had to wait after checking into the emergency room and being seen by a doctor. After collecting data from a random sample of 150 people at the ER, she found a 95% confidence interval for her estimate to be (6.3, 24.1). Which of the following best explains the meaning of her interval?

(A) The true mean amount of time a patient at this local hospital will wait after checking into the emergency room and being seen by a doctor is between 6.3 minutes and 24.1 minutes.

(B) If the true mean waiting time was 26 minutes, then Maria's sample's interval would have been unlikely to have been found.

(C) The true mean amount of time a patient at this local hospital will wait after checking into the emergency room and being seen by a doctor is 15.2 minutes.

(D) If Maria took other samples of 150 random people, then they, too, would yield a 95% confidence interval of (6.3, 24.1).

(E) There is a 95% probability the true mean amount of time a patient at this local hospital will wait after checking into the emergency room and being seen by a doctor is between 6.3 minutes and 24.1 minutes.

19. A manufacturer of batteries has estimated over the years that the mean amount of time their AA batteries last is 1040 minutes of continuous use. However, the company has decided to change some of the materials it uses for its batteries with the hope of making them last longer. Having tested a large sample of batteries, a t-test is carried out using $H_0 : \mu = 1040$ versus $H_a : \mu > 1040$. The t-value for the test is found to be 1.635 and the p-value is found to be 0.052. Using a 5% significance level, which of the following is the correct conclusion for the test?

   (A) Since p > 0.052, we do not have sufficient evidence to conclude that the mean amount of time an AA battery will last is greater than 1040 minutes.

   (B) Since p > 0.052, we do not have sufficient evidence to conclude that the mean amount of time an AA battery will last is less than 1040 minutes.

   (C) Since p > 0.052, we have sufficient evidence to conclude that the mean amount of time an AA battery will last is equal to 1040 minutes.

   (D) Since p > 0.052, we have sufficient evidence to conclude that the mean amount of time an AA battery will last is greater than 1040 minutes.

   (E) Since p > 0.052, we have sufficient evidence to conclude that the mean amount of time an AA battery will last is less than 1040 minutes.

20. Jacob wished to estimate the proportion of people in his town who will vote to support an increase in taxes to secure more funding for the police and fire departments with a 95% confidence interval. How large of a sample would he need to get a margin of error no bigger than 3%?

   (A) 17

   (B) 2,134

   (C) 73

   (D) 11

   (E) 1,068

21. Micah wants to run an experiment to see which leg a professional football player can kick farther with, their dominant leg or non-dominant leg. He elicits 15 professional kickers and has each of them kick a football once with their non-dominant foot and once with their dominant foot. The order of which foot they will kick with first was randomly determined for each player. After he collects his data, which test of significance should he run to determine whether there is a difference between the mean distance kicked between a professional kicker's dominant and non-dominate leg?

    (A) A two-sample *t*-test for the difference between two means

    (B) A two-sample *z*-test for the difference between two proportions

    (C) A matched pairs *t*-test for a mean difference

    (D) A one-sample *t*-test for a proportion

    (E) A one-sample *z*-test for a mean

22. A test was given in a large AP® Biology class. The following boxplot breaks down the scores for the boys and the girls in the class.

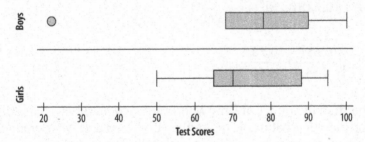

    Which of the following statements must be true?

    (A) The range for the girls is larger than the range for the boys.

    (B) The median score for the girls is larger than the median score for the boys.

    (C) Both sets of data have outliers.

    (D) Fewer girls scored above a 70 on the test than boys who scored above a 70.

    (E) The third quartile for the girls is less than the third quartile for the boys.

23. A national study was conducted in which a random sample of high school-aged students were selected and surveyed about their gender and whether they preferred comfort over style when dressing for school. Of the 234 randomly selected females, 152 preferred comfort over style, and of the 260 randomly selected males, 203 preferred comfort over style. Which of the following is the most appropriate standard error for constructing a confidence interval for the difference in proportions of high school students who prefer comfort over style among males and females?

(A) $\sqrt{\dfrac{(0.5)(0.5)}{234} + \dfrac{(0.5)(0.5)}{260}}$

(B) $\sqrt{\dfrac{(0.72)(0.28)}{234} + \dfrac{(0.72)(0.28)}{260}}$

(C) $\sqrt{\dfrac{(0.65)(0.35)}{234} + \dfrac{(0.78)(0.22)}{260}}$

(D) $\sqrt{(0.72)(0.28)\left(\dfrac{1}{234} + \dfrac{1}{260}\right)}$

(E) $\sqrt{\dfrac{(0.65)^2}{234} + \dfrac{(0.78)^2}{260}}$

24. The student council at Crestwood High School surveyed its student body about what music genre they preferred and what grade they were in. The results are shown in this two-way table:

	Freshmen	Sophomores	Juniors	Seniors	Total
Rap	78	61	54	71	264
Classical	12	9	14	23	58
Rock and Roll	26	36	90	49	201
Top 100	114	100	136	112	462
Country	50	46	12	18	125
Total	280	251	306	273	1110

If a student is selected at random from the student body, what is the probability that the student is a sophomore or prefers rock and roll music?

(A) $\dfrac{36}{1110}$

(B) $\dfrac{36}{251}$

(C) $\dfrac{452}{1110}$

(D) $\dfrac{416}{1110}$

(E) $\dfrac{416}{251}$

25. A group of students at Crestwood High School surveyed its student body about what music genre they preferred and what grade they were in. The results are shown in this two-way table:

	Freshmen	Sophomores	Juniors	Seniors	Total
Rap	78	61	54	71	264
Classical	12	9	14	23	58
Rock and Roll	26	36	90	49	201
Top 100	114	100	136	112	462
Country	50	46	12	18	125
Total	280	251	306	273	1110

If music preference were independent of grade level, which of the following expresses how many students would be expected to be juniors who prefer classical music?

(A) $\dfrac{306 \cdot 1110}{58}$

(B) $\dfrac{251 \cdot 58}{1110}$

(C) $\dfrac{306 \cdot 264}{1110}$

(D) $\dfrac{306 \cdot 58}{1110}$

(E) $\dfrac{58}{1110}$

▌ **Answers and Explanations**

1. **(C)** The addition of a constant to an entire set of data will only affect measures of center, mean, and median. Multiplication of a constant to an entire set of data will affect both measures of center, mean, and median, and measures of spread, range, standard deviation, and IQR. Choice (A) is true because the multiplication of 2.3 and the addition of 5 will both affect the old median. Choice (B) is true because the multiplication of 2.3 and the addition of 5 will both affect the old mean. Choices (D) and (E) are both true because IQR and range are only affected by multiplication. Choice (C) is the correct choice because it is the *incorrect* statement. The standard deviation will only be affected by the multiplication of 2.3 and not the addition of 5. The standard deviation of the new data will be 2.3 times the standard deviation of the old data.

2. **(E)** The sample size is 150 brake pads, 3 boxes with 50 in each. A stratified random sample (A) would select a few brake pads at random *from each* box. A simple random sample (B) would have to allow any possible sample of size 150 the same chance of being selected. Because they are only selecting 3 boxes at random not every group of 150 has a chance. A convenience sample (C) has no randomness to it at all, and this sample did involve randomness. A systematic sample (D) would involve a system that selects every *k*th brake pad, which is not described. A cluster sample (E) begins with creating groups of brake pads based usually on naturally being near each other, hence in boxes. Then, instead of numbering all of the individual brake pads, the manager numbers the groups (or boxes) and selects 3 boxes to be the entire sample. Numbering the boxes 00–99 may indicate a simple random sample, but once a box is chosen, all the brake pads from that box are inspected. So it is impossible to have brake pad 1 chosen and brake pad 2 not chosen, a vital element of an SRS.

3. (A) A sampling distribution for samples of size 50 will have the following mean and standard deviation:

$$\mu_{\bar{x}} = 140 \ grams$$

$$\sigma_{\bar{x}} = \frac{10.6}{\sqrt{50}} \ grams$$

To find the probability that a sample of 50 oranges has a mean above 150 grams will require the z-score for such a sample. The formula for a z-score in a sampling distribution is:

$$z = \frac{\bar{x} - \mu_{\bar{x}}}{\sigma_{\bar{x}}}$$

Choice (B) is looking for the proportion less than 150 grams, not what the question asked. Choice (C) uses the wrong standard deviation for a sample of size 50. Choice (D) forgets the square root on the sample size for the standard deviation and Choice (E) subtracts the sampling distribution mean and the mean in the question in the wrong order.

4. (B) First, in order to interpret slope, you must know which variable is the response (used car price) and which variable is the explanatory (mileage). The easiest way to determine this is to read and see what variable is being predicted (in this case, the used car price). This will always be the response variable. Also, you need to be able to create the regression equation from a computer output. The value next to the word *constant* is the y-intercept; under the word *constant* is the explanatory variable; and the number next to that is the slope. So the regression equation is:

predicted price = 24660.91 – 0.2083(mileage)

Hence, the slope is –0.2083 or approximately –0.21

The generic interpretation of slope is the predicted increase in the response for each increase of 1 unit of the explanatory variable. In this problem, Choice (B) explains slope the best in context. Choice (A) is incorrect because it switches the explanatory and response variable. Choice (C) is incorrect because it uses the y-intercept instead of the slope. Choice (D) is incorrect because it looks at the slope in the reverse order as the predicted increase in the explanatory variable for each increase of 1 unit of the response variable, which is incorrect. Choice (E) is incorrect because it not only flips the explanatory and response variable, but it uses the y-intercept value.

5. (D) By Officer Ron meeting with the selected students, there is a good chance that the students who do use drugs will lie because they are afraid or embarrassed to tell him the truth. Therefore, their response may be untruthful, which is response bias. Choice (A) is incorrect because a non-response bias is when students selected randomly don't respond for some reason. Choice (B) is incorrect because under-coverage bias is when a group of students is left out of being selected, which doesn't seem to have happened. Choice (C) is incorrect because voluntary response bias is when students volunteer instead of being randomly selected, which is not true in this scenario. Choice (E) is incorrect because sampling bias is where the method for selecting the sample is poor or would lead to bias. Officer Ron's selection process is very good, so no sampling bias is present. The issue is how he did the questioning. He should have let students fill out an anonymous survey, where they would be more likely to tell the truth.

6. (E) An element of an experiment is that there has to be one or more treatments. In this case there are two treatments, Razor A and Razor B. Since the problem states that the age of the men does impact the smoothness of their shave, the experiment should block on age. Then in each block it must be randomly decided who gets Razor A and who gets Razor B.

Choice (A) has no randomness to it at all and allows the men to select the razor, which is a very poor design. Choice (B) does block by age, but then gives all men in each age block the same razor, which is incorrect. Choices (C) and (D) attempt a matched pair design, but both are incorrect. Choice (C) pairs up men that are different, for example, the oldest and youngest. A correct matched pair design would contain two men who are as similar as possible. Choice (D) pairs up men correctly by having the two youngest paired together, but then both men in a pair are assigned the same razor, which is incorrect. In a matched pair design, two subjects are paired up because they are very similar and then in each pair both treatments are given out randomly.

Choice (E) allows us to determine the best razor for each age group, youngest, middle, and oldest.

7. (A) This was an observational data collection study. No treatments were given out or imposed on subjects. They simply were measuring for the concentration of nuclear byproduct in the soil, which is the response variable. They did stratify by depth of the soil to see whether the concentration changed due to the depth, but the depth was simply a confounding variable that stratified on. Choice (B) is incorrect because it states that the response variable is the depth of the soil. Again, they were actually looking to measure the concentration of the byproduct. Choices (C) and (D) are incorrect because they claim this was an experiment. Choice (E) is incorrect because a census would require measuring *all* of the soil, which was not done.

8. (B) This is a conditional probability question because you are asked to find the probability of someone getting a stomach virus, given that they have the flu virus.

Let A = getting flu virus

B = getting stomach virus

Then, using the conditional probability formula, one would get:

$$P(A|B) = \frac{P(A \text{ and } B)}{P(B)}$$

$$= \frac{0.08}{0.12}$$

$$= 0.67$$

Choice (A) multiplies the probabilities together, which is the wrong formula and can't be done since the events are not independent. Choice (C) places the probability of getting a stomach virus in the denominator. Choice (D) is the probability of getting a stomach virus without taking into account the given of having the flu virus. Choice (E) is the probability of getting the flu virus or the stomach virus, which is not what the question asks.

9. (C) Let X = the number of pairs of shoes that are sold in one day. First, we need to find the expected value of X. The formula is:

$$E(X) = \mu_X = 0(0.12) + 1(0.20) + 2(0.35) + 3(0.17) + 4(0.10) + 5(0.06)$$

$$= 2.11$$

This means the vendor expects to sell 2.11 pairs of shoes per day. If each pair costs $89, to get the expected daily total dollar amount taken in by the vendor from the sale of shoes, simply multiply 2.11($89) = $187.79.

Choice (A) stops at finding the expected number of shoes sold per day, which isn't even a dollar amount. Choice (B) is the standard deviation of the daily total dollar amount, which is not asked. Choice (D) forgets to take into account the probability of selling 5 shoes. Choice (E) adds up all the shoe outcomes, which is 15, without taking into account their associated probabilities, and then multiplies by $89.

10. (E) To solve this problem, one must first consider that there are three outcomes: 0 survive, 1 survives, or 2 survive. A tree diagram could help. The probabilities of each outcome are below:

$$P(X = 0) = (0.74)(0.85) = 0.6290$$

$$P(X = 1) = (0.74)(0.15) + (0.26)(0.91) = 0.3476$$

$$P(X = 2) = (0.26)(0.09) = 0.0234$$

Choices (A) and (B) do not take into account all possible outcomes. Choices (C) and (D) do not fit the circumstances of the problem.

11. (C) This is a binomial probability scenario in which $n = 4$ and $p = 0.92$. If 3 or 4 of the inspected batteries are in working condition, then the entire day's production will be considered acceptable. So, we must find the probability of 3 in working condition, then all 4 in working condition, then add them together as shown below.

$$P(x = 3) = \binom{4}{3}(0.92^3)(0.08^1) = 0.249$$

$$P(x = 4) = \binom{4}{4}(0.92^4)(0.08^0) = 0.716$$

$$P(x = 3 \text{ or } 4) = 0.249 + 0.716 = 0.965$$

Choice (A) is the probability of only 3 out of the 4 batteries being in working condition. Choice (B) is the probability of all 4 out of the 4 batteries being in working condition. Choice (D) is the probability of the production being found unacceptable, which is 0, 1, or 2 of the 4 inspected batteries being in working condition. Choice (E) is simply the probability that a randomly inspected battery is found to be in working condition.

12. (A) The total standard deviation for the combined weight of the 11 players is found by first getting the variance of one player, which is the standard deviation squared, then multiplying by 11, since each player would be expected to have the same variance. Then take the square root of the combined variance to get the combined weight standard deviation.

$$\sigma_T = \sqrt{15.6^2 \cdot 11} = 51.74$$

Choice (B) simply multiplies 15.6 by the 11 players, which is incorrect. Choice (C) uses the mean weight instead of the standard deviation to calculate the total standard deviation. Choice (C) is the expected total weight of the 11 players, not the standard deviation. Finally, Choice (E) does not find the variance at all. It simply multiplies the standard deviation by 11 and then takes the square root.

13. **(A)** This is a binomial problem in which $n = 10$, $r = 3$, and $p = 0.15$. The formula is given by $p(x = r) = \binom{n}{r} p^r (1-p)^{n-r} = \binom{10}{3}(0.15)^3 (0.85)^7$.

14. **(C)** The null hypothesis is that the true proportion p is still 0.07, while the inspector believes that the true proportion p could now be less than 0.07.

    Choice (A) has the alternative that the true proportion is greater than 0.07, while choice (B) has the alternative that the true proportion is not equal to 0.07, which are both not how the problem is worded. Choice (D) uses the sample proportion in the hypotheses which is never done. Choice (E) has the null and alternative flipped.

15. **(A)** The p-value of 0.042 tells us the probability of the sample mean being as extreme is 0.042, assuming the null hypothesis is true. Since that is less than our significance level, we can conclude that the null hypothesis should be rejected, which means we do have evidence that the alternative hypothesis is true.

    All other choices are a variation of these, but some aspect is incorrect. Choice (B) says the null hypothesis is accepted, which is never concluded. Choice (C) rejects the null hypothesis, but then says there is no evidence that the true mean is not 14. Choice (D) fails to reject the null hypothesis and so does choice (E), which is done if the p-value is greater than the level of significance.

16. **(D)** Since the alternative is that the true mean is greater than 20, we will have the greatest power when the truth is further from the null hypothesis and when the sample size is bigger. Power is the probability of rejecting the null hypothesis and going

with the alternative, when the alternative hypothesis is true. So again, when the truth is further from the null hypothesis , it is clear that the alternative is true, and a bigger sample will always give a more accurate sample mean.

Choices (A) and (B) have a true mean that is close to the null hypothesis, so there is less chance to reject the null hypothesis. Choice (C) has a large mean but a smaller sample size than choice (D). Choice (E) actually has a true mean that is smaller than the null hypothesis, which is against the alternative hypothesis.

Here is another way to think about it. First of all, instead of writing $H_0 : \mu = 20$, you might write it as $H_0 : \mu \leq 20$. That makes sense as either the true average is less than or equal to 20 or the alternative, greater than 20. We want to maximize power which is calculated as $1 - \beta$, where β is the probability of a type II error. Maximizing $1 - \beta$, is the same as minimizing β. So, we want to minimize the chance of rejecting the null hypothesis. Since $\mu = 26$, there is very little chance that the null hypothesis of μ being less than 20 is true, especially since our sample was taken with 500 pieces of data. It would have to be an extraordinary coincidence and is practically impossible. Choice (C) is also remote, but with only 50 pieces of data, it is more likely.

17. **(D)** Due to the skewness and the outliers to the right, the mean and standard deviation will be affected and not be accurate representations of the center and spread. The median and interquartile range are unaffected by skewness and outliers and therefore will give the best representation of the center and spread of the data.

18. **(A)** The correct interpretation is that Maria is 95% confident that the true mean amount of time a patient at this local hospital will wait after checking into the emergency room and being seen by a doctor is between 6.3 minutes and 24.1 minutes.

This exact phrasing is not an option, but there are other correct interpretations for confidence intervals. However, it would also be true that since Maria is so confident that the mean is in her interval, if the true mean was outside of her interval like 26 minutes is, then her interval most likely would not have been found. She most likely would have gotten a sample mean closer to the truth and hence the interval would have captured the truth of 26 minutes.

Choice (A) is close, but does not start with saying that Maria is only 95% confident that the true mean amount of time a patient at this local hospital will wait after checking into the emergency room and being seen by a doctor is between 6.3 minutes and 24.1 minutes. Choice (C) says the true mean is the center of her interval, where the center of her interval 15.2 is only her sample mean, not the true mean. Choice (D) is incorrect because another sample will most likely yield a different confidence interval and not the exact same interval as Maria's. Choice (E) claims there is a 95% probability. The 95% represents a confidence level, not a probability.

19. (A) Since the p-value is greater than the indicated level of significance, the conclusion will be to fail to reject the null, meaning there is not enough evidence to claim that the true mean battery life is greater than 1040 minutes.

    Choice (B) states less than 1040 minutes, which is not what our alternative hypothesis was. Choice (C) states that we do have evidence that the mean amount of time the batteries will last is equal to 1040 minutes. We do not have evidence to reject the null, but that does not mean that the null is still true. Choices (D) and (E) state that we do have sufficient evidence, which is incorrect.

20. (E) The margin of error that this problem is looking for is 0.03 with 95% confidence. 95% confidence requires a z-score of 1.96. Since we do not know the sample proportion p^*, nor is one mentioned in the problem, the safest choice is $p^* = 0.5$. To get the correct answer, the following inequality needs to be solved.

$$1.96\sqrt{\frac{p^*(1-p^*)}{n}} < 0.03$$

$$1.96\sqrt{\frac{0.5 \times 0.5}{n}} < 0.03$$

$$\sqrt{\frac{0.25}{n}} < \frac{0.03}{1.96}$$

$$\frac{0.25}{n} < 0.153^2$$

$$n > \frac{0.25}{0.153^2}$$

$$n > 1067$$

Since n is an integer, $n = 1068$

21. **(C)** The key is that in this problem there was only one sample (not two, so Choices (A) and (B) are incorrect). The sample of 15 kickers was simply measured twice, so each kicker was matched with himself, and the difference between the distance of the kick with his dominant foot and non-dominant foot was calculated. This makes this a matched pairs test for the mean difference. Choices (B) and (D) are also incorrect because this problem had nothing to do with testing proportions.

22. **(E)** The 3rd quartile of the girls looks to be around 88 points while the boys' 3rd quartile is around 90 points. The range of a set of data does include outliers, so the boys have a much larger range, making Choice (A) false. The median for the girls is around 70, while the boys' median is around 79. Thus, Choice (B) is false. Only the boys' data have an outlier marked, so Choice (C) is false. For the girls, around 50% scored more than 70, and for the boys one can clearly tell that more than 50% scored above 70. But without knowing how many boys and girls were in the survey, it is impossible to conclude that a smaller number of girls scored above 70. If there are more girls than boys in the survey, then even though a smaller proportion

of girls scored above 70, it could still equate to more girls than boys. Choice (D) could possibly be true, but it isn't a sure thing.

23. **(C)** In finding the standard error for a confidence interval, there is no assumption that the two populations should have the same proportions; hence, we would not pool the data together. The correct formula for the standard error for the difference between two population proportions is:

$$\text{S.E.} = \sqrt{\frac{\hat{p}_1(1-\hat{p}_1)}{n_1} + \frac{\hat{p}_2(1-\hat{p}_2)}{n_2}}$$

$$\hat{p}_1 = \frac{152}{234} \approx 0.65$$

$$\hat{p}_2 = \frac{203}{260} \approx 0.78$$

Choice (A) uses a generic 0.5 for both sample proportions. Choice (B) or Choice (D) would be correct if we were to pool the data together for a test of significance, but not for a confidence interval. Choice (E) does not use the proper formula; rather it is using the formula for the standard error for the difference between two means, but this is a proportion problem.

24. **(D)** To arrive at this answer one would need to apply the addition rule for probability:

$$P(A \text{ or } B) = P(A) + P(B) - P(A \text{ and } B)$$

$$P\left(\begin{array}{c}\text{sophomore} \\ \text{or rock/roll}\end{array}\right) = P(\text{sophomore}) + P(\text{rock/roll}) - P\left(\begin{array}{c}\text{sophomore} \\ \text{or rock/roll}\end{array}\right)$$

$$P\left(\begin{array}{c}\text{sophomore} \\ \text{or rock/roll}\end{array}\right) = \frac{251}{1110} + \frac{201}{1110} - \frac{36}{1110}$$

$$P\left(\begin{array}{c}\text{sophomore} \\ \text{or rock/roll}\end{array}\right) = \frac{416}{1110}$$

Choice (A) is the probability of selecting a sophomore *and* someone who prefers rock and roll. Choice (B) is the conditional probability of selecting someone who prefers rock and roll music, given that the student is a sophomore. Choice (C) does not subtract from the overlap of the students who are *both* sophomores and prefer rock and roll music. Without subtracting them, those 36 students will get counted twice. Choice (E) has the correct numerator, but the denominator only counts the sophomores. However, we are selecting from the entire student body. Since it is greater than one, it cannot be correct.

25. (D) If grade level and music preference are independent, then the proportion of students who prefer classical music will be the same proportion on each individual grade level. We know that 58/1110 is the proportion of students who prefer classical music. If the variables are independent, then that same proportion of juniors will prefer classical music. Hence, 306 juniors (58/1110) will prefer classical music. So the answer is $\frac{306 \cdot 58}{1110}$.

Some teachers may also teach the formula to find an expected value by taking the row total multiplied by the column total divided by the grand total. That would work as well to arrive at a solution of $\frac{306 \cdot 58}{1110}$.

Choice (A) has the row total in the dominator. Choice (B) shows the expected number of sophomores, not juniors. Choice (C) shows the juniors we expect to prefer rap music. Choice (E) is only the total proportion of all students that prefer classical music. All of these choices are incorrect.

Practice Free-Response Questions

Practice with the following AP®-style questions. Then go online to access our timed, full-length practice exam at *www.rea.com/ studycenter.*

1. Many companies make claims about their products in order to corner a larger share of the market. Consumer advocacy groups conduct tests of company products and rate their overall effectiveness so consumers can make wise choices. Company A claims to produce a weight-loss drug (Drug A) that assists people in losing weight more effectively with a balanced diet and exercise plan. Company B also produces a weight-loss drug (Drug B) that also claims to help people lose weight.

   An independent consumer advocacy group conducted the following experiment. Forty people (20 men and 20 women) of similar weight and obesity levels were randomly split into 2 groups of 10 men and 10 women. Members of the two groups were given the drug from company A or company B. The study was done over a 4-week period and all group members were assigned the same diet and exercise routine over the 4 weeks. At the end of the four weeks, each person's weight loss was measured simply by finding his or her weight difference for the 4-week period.

   The following table summarizes the calculated differences. A negative number indicates that the subject actually gained weight while on the drug.

Drug	Values Below Q₁	Q₁	Median	Q₃	Values Above Q₃
A	–4, –3, –1	1	2	5	6, 8, 14
B	–1, 2, 3	5	14	20	22, 24, 25

(a) On the grid below, display parallel boxplots (showing outliers, if any) of the differences between the two drugs.

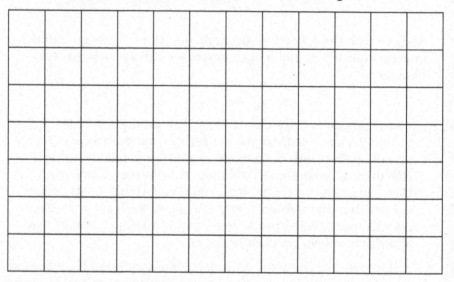

(b) Which drug would you recommend has a better efficacy on weight loss?

_____

(c) Using a mathematical method, show which (if any) of the data points above are statistical outliers.

_____

(d) Company C also claims that it can produce a weight-loss drug with more significant results. It claims the drug is gender specific and works best for females. Describe an experiment that can be done to test this claim.

_____

_____

_____

_____

2. A political watchdog group based in Washington, D.C., has concluded that certain members of the House of Representatives are not voting on legislation as often as they should. There are currently 435 members of the House and findings were based on archived data of 40 random representatives. The data set below shows the number of votes missed by these 40 random representatives.

11	13	31	7	24	17	11	22	6	14
17	6	54	22	25	14	9	19	21	27
23	5	24	17	22	4	13	15	14	11
22	44	22	13	11	20	21	19	15	14

(a) Display the data shown above in a stemplot.

(b) Using the stemplot you created, describe the main features of this distribution.

_____

_____

_____

_____

(c) Determine the mean, median, mode, and standard deviation.

_____

_____

_____

_____

3. A basketball coach is figuring out the best shooting distance for each of two players (Player 1 and Player 2) who play the center position on the team. To determine the optimal distance, he records the number of feet away from the basket each player is when he successfully makes a shot. Each player takes 40 shots, but only successful attempts are recorded. The distances for successful shots are as follows:

Player 1	2.0	4.0	5.5	3.0	1.0	2.0	3.5	4.0	2.5	3.0	2.0	1.0	1.5	2.0	2.0
Player 2	2.0	6.5	4.5	7.0	5.5	6.0.	2.0	4.0	6.0	7.5	6.0	4.0	X	X	X

(a) Construct a parallel dotplot of the data and describe the center, shape, and spread.

(b) On average which player tends to be a better shooter from a longer distance? Explain.

_____

_____

_____

_____

(c) The coach decides that he wants to use crude laws of probability to decide who will take the game-winning shot at the end of a game. Using simple laws of probability, how would you advise the coach?

_____

_____

_____

_____

4. Fatal accidents on the road happen all year round and are dependent on weather, type of vehicle, and driving age/experience. In 2004, the number of fatal deaths in car accidents was recorded in the state of Wyoming. The age of the driver was recorded in each of these fatal accidents and placed in a table as shown below.

    (a) Fill in the missing parts of the table.

Age at death	Frequency	Relative Frequency	Cumulative Relative Frequency
10 to <15	2		
15 to <20	3		
20 to <25	8		
25 to <40	11		
40 to <55	9		
55 to <70	8		
>70	3		

    (b) Construct a cumulative frequency plot for the data using the space below.

5. Because of economic restructuring, employees of a company were terminated so the company could stay in business. Managers presented an analysis of terminated employees to the human resources department to ensure that an unbiased termination plan had been implemented. Before the dismissals, 2,000 employees worked for the company.

Table A below shows, by gender, the number of employees who were retained or fired.

Table A

Gender	Retained	Fired	Total
Male	600	400	1000
Female	800	200	1000
Total	1400	600	2000

Legal counsel for the human resources department was dissatisfied with the analysis. They were concerned that more men than women were terminated which could lead to a lawsuit. The lawyers asked for a more detailed analysis based on the positions each member had with the company.

Table B below represents the breakdown of fired and retained employees according to the position they held at the company.

Table B

Marketing Position	Retained	Fired	Total
Male	510	390	900
Female	50	50	100
Total	560	440	1000
Financial Position	**Retained**	**Fired**	**Total**
Male	90	10	100
Female	750	150	900
Total	840	160	1000

(a) Calculate the marginal frequency of men and women who were retained or fired.

_____

_____

_____

_____

(b) Using the graph below, display the data in table B.

(c) According to the data, men tended to be fired "more frequently" than women. Comment on whether you believe that the company's legal counsel should be concerned about the possibility of a successful lawsuit.

_____

_____

_____

_____

6. A company that manufactures laundry detergent believes the effectiveness of its detergent depends on the concentration of Chemical X used in the detergent. A study of the effectiveness of Chemical X is conducted. 15 white t-shirts are dipped into a pan that contains mud from the Mississippi River. The t-shirts are then washed in separate, but the same type of, washing machines with detergent that contains various concentrations of Chemical X. The effectiveness of the detergent is then measured by digital analysis of the shade of mud (a percentage) that is left on the t-shirts. A high percentage of shade indicates that mud was retained on the shirt. The data for the 15 t-shirts is summarized in the table below.

Amount of Chemical X (mg) in Detergent	Percentage of Shade
1.0	2.0
1.2	3.3
1.4	5.0
1.6	6.7
1.8	8.0
2.0	9.9
2.2	11.1
2.4	12.7

(continued)

(continued from previous page)

Amount of Chemical X (mg) in Detergent	Percentage of Shade
2.6	14.5
2.8	16.0
3.0	17.3
3.2	18.9
3.4	20.5
3.6	22.0
3.8	23.7

(a) Using the graph below, construct a scatterplot of the data above.

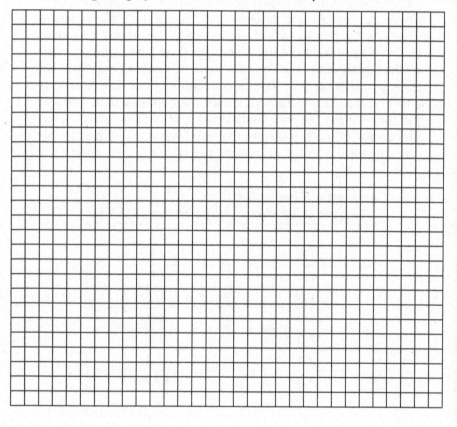

(b) Hand-draw the line of best fit on the graph and find the least-squares regression equation.

(c) What does the slope of the mean indicate in this study?

_____

_____

_____

_____

(d) Using your calculator, find the least-squares regression line. How does that line compare to your "hand drawn" line of best fit?

_____

_____

_____

_____

(e) Using the least-squares regression line from your calculator, calculate the residual for 3.0 mg of detergent added.

_____

_____

_____

_____

(f) Comment on both the correlation coefficient and coefficient of determination.

_____

_____

_____

_____

7. Health officials at a local hospital are trying to determine how to decrease the amount of blood clots in individuals who suffer from diabetes. One physician believes that if patients take doses of iron that are above the required daily recommended amount, the incidence of blood clots over a designated period will decrease. To test this hypothesis, diabetics are monitored for daily iron intake and the number of blood clots is measured over a 60-day period. Below are two plots showing the data. Plot 1 is a least-squares regression plot of the number of blood clots versus the average amount of iron intake. A negative number indicates the patient took less iron on average than the daily recommended amount. Plot 2 is a residual plot of the data.

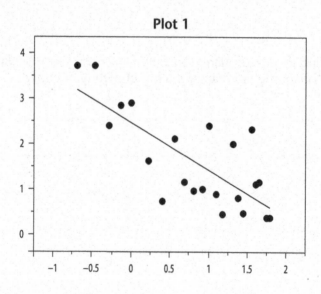

Plot 1

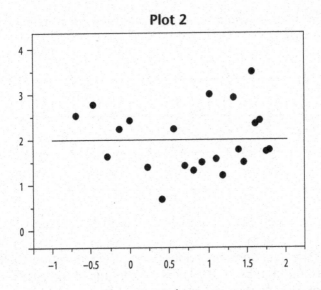

Plot 2

(a) The physician did not provide the correct axis labels for the plots. Provide the correct labels for both plots.

_____

_____

_____

_____

_____

(b) Comment on whether you believe there is a correlation from the data.

_____

_____

_____

_____

_____

(c) For the data on the previous pages, the correlation coefficient was found to be –0.81. What percentage of the data can be explained by the least-squares equation?

_____

_____

_____

_____

_____

8. In 2005, after Hurricane Katrina devastated the greater New Orleans, Louisiana, area, a pest control problem developed. In the months after the storm, the area remained wet and damp, allowing for the breeding of mosquitoes, ants, and bees. To control the pest problem, areas of New Orleans were chemically treated with a pesticide called "Bug Buster" and mortality rates of the pests were measured. A 22% mortality rate indicates that the pesticide killed 22% of the pests in the treated area. The following data summarizes the mortality rates:

Concentration of Bug Buster (percentage)	Number of Exposed Pests	Mortality Rate
2	245	0.05
4	310	0.33
6	275	0.40
8	281	0.51
10	291	0.63
12	230	0.80
14	333	0.86
16	299	0.87
18	262	0.89
20	271	0.91

(a) Using the graph below, construct a scatterplot of the data above.

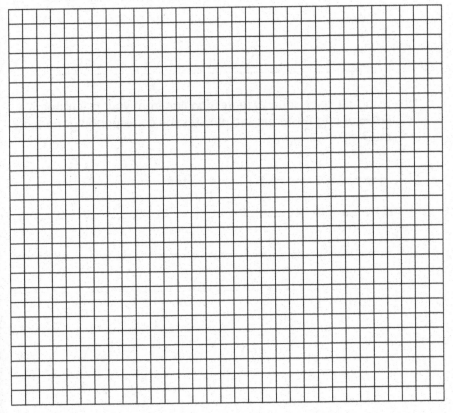

(b) Does a simple linear relationship seem to be a good fit? Explain why.

_____

_____

_____

_____

(c) Transform the data appropriately and redefine the least-squares regression equation.

_____

_____

_____

_____

(d) Using the least-squares regression line from your calculator, if 6.4% of Bug Buster is used, what is the corresponding mortality rate?

_____

_____

_____

_____

9. Researchers are trying to investigate the link between breast cancer and workplace stress. Researchers asked 452 women suffering from breast cancer to classify their stress in the workplace as low, medium, or high. A control group of women with no history of breast cancer was also asked to assess their workplace stress level. Below is the data:

Patient Category	Workplace Stress-Level Rating		
	Low	Medium	High
Breast Cancer Patient	152	202	98
Non-Breast-Cancer Patient	134	123	56

(a) Identify the observational units in the study.

_____

_____

_____

_____

(b) Is this a controlled experiment or an observation?

_____

_____

_____

_____

(c) To the nearest tenth of a percent, calculate the marginal frequencies for each category.

_____

_____

_____

_____

10. A horticulturist is measuring the effects of soil composition and temperature on the growth rate of vinca plants. The experiment is conducted in a controlled environment where 10 vinca plants are randomly placed in similar troughs of soil. The horticulturist is planning to use 3 different types of soil (1, 2, 3) and two different temperatures.

(a) List the treatments in this experiment.

_____

_____

_____

_____

(b) Using the treatments in part (a), describe a completely randomized design to compare vinca growth.

_____

_____

_____

_____

(c) Describe one advantage and one disadvantage based on statistics to only using vinca as the experimental unit.

_____

_____

_____

_____

11. At large stadiums during the summer, beverage companies fill soda cups with ice and vendors take them through the stadium to sell. The ice melts and customers end up with watered-down drinks. People wonder whether these companies intentionally put too much ice in the drink. Following is a random sample of the amount of liquid in 30 16-ounce cups provided by the company, Shivers.

8.8	8.7	9.3	10.6	8.4	8.5
9.4	8.4	9.0	9.7	8.9	10.2
10.3	9.6	10.3	9.4	9.7	9.0
10.4	11.0	9.5	8.6	8.3	8.9
10.9	10.2	9.0	8.5	8.0	10.5

(a) Find the 95% confidence interval for the data and write a complete sentence to interpret its meaning.

_____

_____

_____

_____

(b) Mr. Adams complains to the manager of Shivers that there was too much ice in his 16-ounce drink. The manager explains that Mr. Dawson was just unlucky, and that Icey Beverage usually puts 10 ounces in a drink. Comment on whether the manager was truthful.

_____

_____

_____

_____

(c) Explain whether there is evidence that Shivers puts less than 9.8 ounces of soda in their drinks (meaning too much ice). Show important information and your conclusion.

_____

_____

_____

_____

12. How long will it take customers to switch from a slow-moving supermarket check-out line to another? An experiment was set up on a busy day at a market. A checkout line was set up so that the cashier worked very slowly. Statistics were kept about whether people switched to another checkout line. Eighty-two of 245 men changed lines compared to 55 of 215 women.

(a) Test the premise at the 5% level that men and women make different decisions about changing lines.

_____

_____

_____

_____

(b) What is the standard error of the difference in proportion of men and women who switch lines?

_____

_____

_____

_____

(c) Find the 95% confidence interval in the difference in proportion of men and women who switch lines.

_____

_____

_____

_____

13. 140 randomly selected women in their 20s were asked about their favorite place to go on a Saturday night.

The results are in the table below.

Movies	Sports Event	Dining Out	Shopping	Concert
29	21	39	30	21

(a) Test the claim that each choice is equally as popular.

_____

_____

_____

_____

(b) At the same time, a random sample of 20-year-old men were asked about their favorite place to go on a Saturday night. The results are in the table below. Explain what the null hypothesis H_0 says about gender.

Movies	Sports Event	Dining Out	Shopping	Concert
28	40	29	12	25

_____

_____

_____

_____

(c) Find the expected counts and display them in a two-way table (one decimal place).

_____

_____

_____

_____

(d) Perform the appropriate test for the two sets of data.

_____

_____

_____

_____

Answers and Explanations

1. (a)

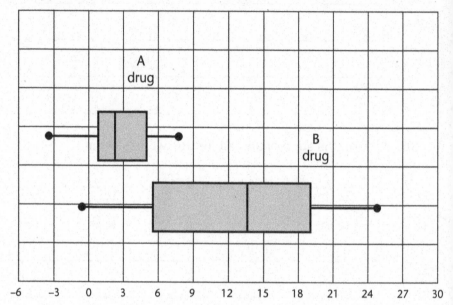

(b) Drug B is the better group. The median weight loss was 14 pounds, very few people gained weight, and the mean weight loss was higher.

(c) Using the 1.5 × IQR rule, find the following:

Drug A

$1.5 \times 4 = 6$

$Q_1 - 6 = -5$

$6 + Q_3 = 11$

Therefore, 14 is an outlier.

Drug B

$1.5 \times 15 = 22.5$

$Q_1 - 22.5 = -17.5$

$22.5 + Q_3 = 42.5$

There are no outliers.

(d) The drug will be randomly administered to blocked groups of males and females. The results could be analyzed using boxplots and summary statistics. A control group for each block of placebo would indicate if Company C's drug would have any efficacy.

2. (a)

```
0 | 4  5  6  6  7  9
1 | 1  1  1  1  3  3  3  4  4  4  4  5  5  7  7  7  9  9
2 | 0  1  1  2  2  2  2  2  3  4  4  5  7
3 | 1
4 | 4
5 | 4
```

Stem   Leaf

(b) The data is skewed slightly to the right, with a good portion of it clustered from 10–20.

(c)

Mean	18.0
Median	17.0
Mode	22.0
St. Dev	9.7

3.

```
  •                                                 •
  •                     •                           •
  •                     •    •          •    •    •    •    •   Player 2
_____

                       •
                       •
                       •
  •                    •         •          •
  •    •    •    •    •    •    •              •                Player 1
_____
 1.0  1.5  2.0  2.5  3.0  3.5  4.0  4.5  5.0  5.5  6.0  6.5  7.0  7.5
```

(a) Center—Player 1 center is around 2–2.5 while player 2 is 5.5–6.

Shape—Players 1 and 2 are somewhat symmetric with possible outliers.

Spread—Player 1 has a better range than Player 2.

(b) Player 2 is a better distance shooter since his shots are plotted more toward the higher values. Player 1's average is 2.5, while player 2's average is 5.1.

(c) Player 1 makes 15 of 40 shots, while player 2 makes 12 shots out of 40. Based on probability, player 1 has a better chance of successfully making the shot.

4. (a)

Age at death	Frequency	Relative Frequency	Cumulative Relative Frequency
10 to <15	2	0.045	0.045
15 to <20	3	0.068	0.113
20 to <25	8	0.182	0.295
25 to <40	11	0.250	0.545
40 to <55	9	0.205	0.75
55 to < 70	8	0.182	0.932
>70	3	0.068	1.00

(b)

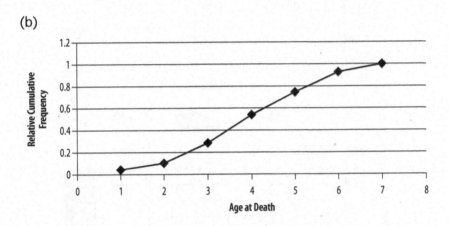

1 = 10 to <15
2 = 15 to <20
3 = 20 to <25
4 = 25 to <40
5 = 40 to <55
6 = 55 to <70
7 = >70

5. (a)

Gender	Retained	Fired
Male	60%	40%
Female	80%	20%

(b)

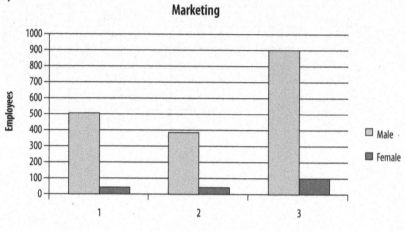

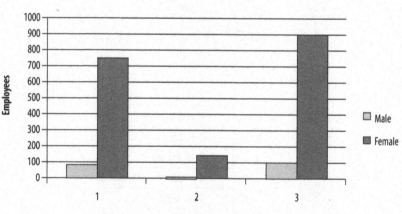

(c) This is an example of Simpson's Paradox. Since there were more males from marketing who were fired (because they made up most of the marketing team), it seems as if men were discriminated against.

6. (a & b)

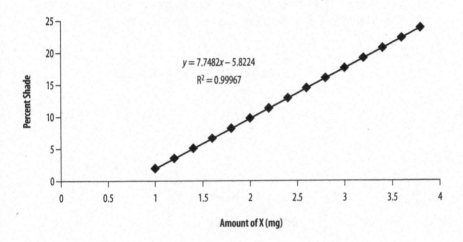

Percent Shade vs. Detergent Amount

$y = 7.7482x - 5.8224$
$R^2 = 0.99967$

Percent Shade (y-axis)
Amount of X (mg) (x-axis)

(c) Slope means that for every 1 mg of Chemical X used the percent shade rises 7.7482%.

(d) Varying answers, but a hand-drawn graph is not better than one on a calculator or statistical program.

(e) $\hat{y} = 7.748x - 5.8224$

$\hat{y} = 7.7482(3) - 5.8224 = 17.42$

residual $= y - \hat{y}$

residual $= 17.3 - 17.42 = -0.12$

(f) $r = 0.99$ indicating a strong positive correlation.

$R^2 = 99.9\%$ indicating the least squares equations explains 99.9% of the variability in the data.

7. (a) The axes should be labeled as:

   Number of blood clots (*y*-axis)

   Average amount of iron intake (*x*-axis).

   (b) There is a negative correlation based on the least squares lines drawn.

   (c) Since $r = 0.81$ then $R^2 \approx 66\%$, so approximately 66% of the variability in the data can be explained by this equation.

8. (a)

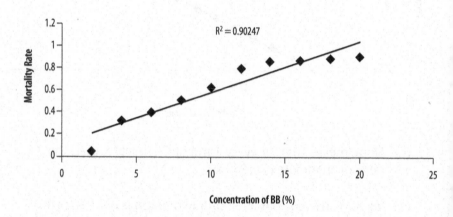

Mortality Rate vs. Concentration BB

$R^2 = 0.90247$

   (b) Since the line is curved, it does not indicate a good linear fit.

(c)

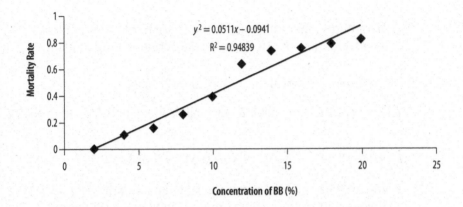

Mortality Rate Squared vs. Concentration

$$y^2 = 0.0511x - 0.0941$$
$$R^2 = 0.94839$$

R² increased to 94% by squaring the mortality rate as the transformation.

(d) $\hat{y}^2 = 0.0511x - 0.0941$

$y^2 = 0.0511(6.4) - 0.0941$

$y^2 = 0.23$

$\hat{y} = 48\%$

9. (a) The experimental units are the women in the workplace.

(b) Observation—no treatments have been tested.

(c)

Category	Low	Medium	High
Breast Cancer Patient	33.6%	44.7%	21.7%
Non-Breast-Cancer Patient	42.8%	39.3%	17.9%

10. (a) Treatments—soil and temperature

    (b) Randomize each of the three soils with all the temperatures being tested.

    (c) Advantage—less variability in the study to block or control. Disadvantage—the study only relates to one type of plant.

11. 95% confidence interval for $\mu = x \pm 2.042\left(\dfrac{s}{\sqrt{n}}\right)$

    (2.42 comes from Table B with tail probability 0.25 and df = 29)

    (a) $9.40 \pm 2.042\left(\dfrac{0.848}{\sqrt{30}}\right) = 9.40 \pm 0.317 = 9.083$ to 9.717

    We are 95% confident that the average amount of liquid Shivers puts in their 16-oz. drinks is between 9.083 and 9.717 ounces.

    (b) Since 10 is not within the 95% confidence interval of 9.083 to 9.717, we are fairly confident 10 ounces is not the average amount of ice in a cup of soda and the manager is not telling the truth. The sample size of 30 is big enough to feel confident in this opinion.

    (c) One-sample *t*-test

    $H_0: \mu = 9.8$

    $H_a: \mu < 9.8$

    $t = \dfrac{\bar{x} - \mu}{\frac{s}{\sqrt{n}}} = \dfrac{9.4 - 9.8}{\frac{0.848}{\sqrt{30}}} = 2.584\mu$ $p = 0.008$

    Reject H_0 and accept H_a. There is good evidence that Shivers puts less than 9.8 ounces of soda in their cups (and thus too much ice).

12. (a) Two-sample *z*-proportion test

    $n_M = 245, x_M = 82, n_W = 215, x_W = 55$

    $H_0: p_M - p_W = 0$ or $pM = p_W$

H_a: $p_M - p_W \neq 0$ or $pM \neq p_W$

$z = 1.846$ $p = 0.061$ (by calculator)

Fail to reject H_0 so there is not enough evidence at the 5% level that men and women make different decisions as to whether they will switch lines.

(b) $SE = \sqrt{\dfrac{\left(\dfrac{82}{245}\right)\left(\dfrac{163}{245}\right)}{245} + \dfrac{\left(\dfrac{55}{215}\right)\left(\dfrac{160}{215}\right)}{215}} = 0.042$

(c) $\left(\dfrac{82}{245} - \dfrac{55}{215}\right) \pm 1.96(0.042) = 0.079 \pm 0.082$

$$= -0.004 \text{ to } 0.162$$

We are 95% confident that the difference in proportion of men and women switching lines is between –0.4% and 16.2. This means that there is a small chance that a greater percentage of women will change lines.

13. (a) H_0: Women prefer all activities on a Saturday night equally.

H_a: Women don't prefer all activities on a Saturday night equally.

$\chi^2 = 8.00$,   df = 4 $p = 0.092$ so fail to reject H_0

There is no evidence that there is a difference in women's preferences on a Saturday night.

(b) H_0: What 20 year-olds want to do on a Saturday night is independent of gender.

(c)

	Movies	Sports Event	Dining Out	Shopping	Concert
Women	29.1	31.2	34.7	21.4	23.5
Men	27.9	29.8	33.3	20.5	22.5

(d) H_0: What people want to do on Saturday night is independent of gender.

H_a: What people want to do on Saturday night is not independent of gender. $\chi^2 = 15.34$, df $= 4$ $p = 0.004$ so reject H_0, accept H_a

All expected counts are at least 5.

Evidence that gender matters in what 20-year-olds want to do on a Saturday night.

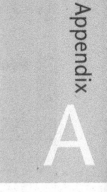

Formulas and Tables

In statistics, there are commonly used variables for different measurements. In many problems, the measurement and its variable will usually be identified. But in computer-generated output, typically only the variable is used. So, for example, when summary statistics is given with $s = 4$, students should know that this refers to the standard deviation of the sample.

I. Descriptive Statistics

Formula	Explanation
$\bar{x} = \dfrac{\sum x_i}{n}$	Average or arithmetic mean
$s_x = \sqrt{\dfrac{1}{n-1}\sum(x_i - \bar{x})^2}$	Standard deviation
$s_p = \sqrt{\dfrac{(n_1 - 1)s_1^2 + (n_2 - 1)s_2^2}{(n_1 - 1) + (n_2 - 1)}}$	Pooled sample standard error
$\hat{y} = b_0 + b_1 x$	Simple linear regression line
$b_1 = \dfrac{\sum(x_i - \bar{x})(y_i - \bar{y})}{\sum(x_i - \bar{x})^2}$	Regression coefficient (slope)
$b_0 = \bar{y} - b_1\bar{x}$	Regression slope intercept
$r = \dfrac{1}{n-1}\sum\left(\dfrac{x_i - \bar{x}}{s_x}\right)\left(\dfrac{y_i - \bar{y}}{s_y}\right)$	Sample correlation coefficient
$b_1 = r\dfrac{s_y}{s_x}$	Regression coefficient (slope)
$s_{b_1} = \dfrac{\sqrt{\dfrac{\sum(y_i - \hat{y}_i)^2}{n-2}}}{\sqrt{\sum(x_i - \bar{x})^2}}$	Standard error of the regression slope

II. Probability

Formula	Explanation
$P(A \cup B) = P(A) + P(B) - P(A \cap B)$	Rule of addition
$P(A \mid B) = \dfrac{P(A \cap B)}{P(B)}$	Rule of multiplication
$E(X) = \mu_x = \sum x_i p_i$	Expected value of X
$Var(X) = \sigma_x^2 = \sum (x_i - \mu_x)^2 p_i$	Variance of X

If X has a binomial distribution with parameters n and p, then:

Formula	Explanation
$P(X = k) = \dbinom{n}{k} p^k (1-p)^{n-k}$	Binomial formula
$\mu_x = np$	Binomial distribution
$\sigma_x = \sqrt{np(1-p)}$	Standard deviation of binomial distribution
$\mu_{\hat{p}} = p$	Mean of sampling distribution of the proportion
$\sigma_{\hat{p}} = \sqrt{\dfrac{p(1-p)}{n}}$	Standard deviation of the sampling distribution of the proportion

If $\bar{x}$ is the mean of a random sample of size n from an infinite population with mean m and standard deviation σ, then:

Formula	Explanation
$\mu_x = \mu$	Mean of sampling distribution of the proportion
$\sigma_x = \dfrac{\sigma}{\sqrt{n}}$	Standard deviation of the sampling distribution of the mean

Inferential Statistics

$$\text{Standardized test statistic} = \frac{\text{statistic} - \text{parameter}}{\text{standard deviation of statistic}}$$

Confidence interval = statistic ± (critical value) · (standard deviation of statistic).

Single-Sample

Statistic	Standard Deviation of Statistic
Sample mean	$\dfrac{\sigma}{\sqrt{n}}$
Sample proportion	$\sqrt{\dfrac{p(1-p)}{n}}$

Two-Sample

Statistic	Standard Deviation of Statistic
Difference of sample means	$\sqrt{\dfrac{\sigma_1^2}{n_1} + \dfrac{\sigma_2^2}{n_2}}$ Special case when $\sigma_1 = \sigma_2$ $\sigma\sqrt{\dfrac{1}{n_1} + \dfrac{1}{n_2}}$
Difference of sample proportions	$\sqrt{\dfrac{p_1(1-p_1)}{n_1} + \dfrac{p_2(1-p_2)}{n_2}}$ Special case when $p_1 = p_2$ $\sqrt{p(1-p)}\sqrt{\dfrac{1}{n_1} + \dfrac{1}{n_2}}$

$$\text{Chi-square test statistic} = \sum \frac{(\text{observed} - \text{expected})^2}{\text{expected}}$$

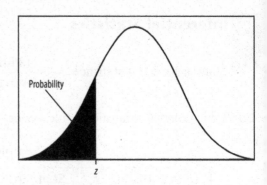

Table entry for *z* is the
probability lying below *z*.

Table A   Standard Normal Probabilities

Z	0.00	0.01	0.02	0.03	0.04	0.05	0.06	0.07	0.08	0.09
-3.4	0.0003	0.0003	0.0003	0.0003	0.0003	0.0003	0.0003	0.0003	0.0003	0.0002
-3.3	0.0005	0.0005	0.0005	0.0004	0.0004	0.0004	0.0004	0.0004	0.0004	0.0003
-3.2	0.0007	0.0007	0.0006	0.0006	0.0006	0.0006	0.0006	0.0005	0.0005	0.0005
-3.1	0.0010	0.0009	0.0009	0.0009	0.0008	0.0008	0.0008	0.0008	0.0007	0.0007
-3.0	0.0013	0.0013	0.0013	0.0012	0.0012	0.0011	0.0011	0.0011	0.0010	0.0010
-2.9	0.0019	0.0018	0.0018	0.0017	0.0016	0.0016	0.0015	0.0015	0.0014	0.0014
-2.8	0.0026	0.0025	0.0024	0.0023	0.0023	0.0022	0.0021	0.0021	0.0020	0.0019
-2.7	0.0035	0.0034	0.0033	0.0032	0.0031	0.0030	0.0029	0.0028	0.0027	0.0026
-2.6	0.0047	0.0045	0.0044	0.0043	0.0041	0.0040	0.0039	0.0038	0.0037	0.0036
-2.5	0.0062	0.0060	0.0059	0.0057	0.0055	0.0054	0.0052	0.0051	0.0049	0.0048
-2.4	0.0082	0.0080	0.0078	0.0075	0.0073	0.0071	0.0069	0.0068	0.0066	0.0064
-2.3	0.0107	0.0104	0.0102	0.0099	0.0096	0.0094	0.0091	0.0089	0.0087	0.0084
-2.2	0.0139	0.0136	0.0132	0.0129	0.0125	0.0122	0.0119	0.0116	0.0113	0.0110
-2.1	0.0179	0.0174	0.0170	0.0166	0.0162	0.0158	0.0154	0.0150	0.0146	0.0143
-2.0	0.0228	0.0222	0.0217	0.0212	0.0207	0.0202	0.0197	0.0192	0.0188	0.0183
-1.9	0.0287	0.0281	0.0274	0.0268	0.0262	0.0256	0.0250	0.0244	0.0239	0.0233
-1.8	0.0359	0.0351	0.0344	0.0336	0.0329	0.0322	0.0314	0.0307	0.0301	0.0294
-1.7	0.0446	0.0436	0.0427	0.0418	0.0409	0.0401	0.0392	0.0384	0.0375	0.0367
-1.6	0.0548	0.0537	0.0526	0.0516	0.0505	0.0495	0.0485	0.0475	0.0465	0.0455
-1.5	0.0668	0.0655	0.0643	0.0630	0.0618	0.0606	0.0594	0.0582	0.0571	0.0559
-1.4	0.0808	0.0793	0.0778	0.0764	0.0749	0.0735	0.0721	0.0708	0.0694	0.0681
-1.3	0.0968	0.0951	0.0934	0.0918	0.0901	0.0885	0.0869	0.0853	0.0838	0.0823
-1.2	0.1151	0.1131	0.1112	0.1093	0.1075	0.1056	0.1038	0.1020	0.1003	0.0985
-1.1	0.1357	0.1335	0.1314	0.1292	0.1271	0.1251	0.1230	0.1210	0.1190	0.1170
-1.0	0.1587	0.1562	0.1539	0.1515	0.1492	0.1469	0.1446	0.1423	0.1401	0.1379
-0.9	0.1841	0.1814	0.1788	0.1762	0.1736	0.1711	0.1685	0.1660	0.1635	0.1611
-0.8	0.2119	0.2090	0.2061	0.2033	0.2005	0.1977	0.1949	0.1922	0.1894	0.1867
-0.7	0.2420	0.2389	0.2358	0.2327	0.2296	0.2266	0.2236	0.2206	0.2177	0.2148
-0.6	0.2743	0.2709	0.2676	0.2643	0.2611	0.2578	0.2546	0.2514	0.2483	0.2451
-0.5	0.3085	0.3050	0.3015	0.2981	0.2946	0.2912	0.2877	0.2843	0.2810	0.2776
-0.4	0.3446	0.3409	0.3372	0.3336	0.3300	0.3264	0.3228	0.3192	0.3156	0.3121
-0.3	0.3821	0.3783	0.3745	0.3707	0.3669	0.3632	0.3594	0.3557	0.3520	0.3483
-0.2	0.4207	0.4168	0.4129	0.4090	0.4052	0.4013	0.3974	0.3936	0.3897	0.3859
-0.1	0.4602	0.4562	0.4522	0.4483	0.4443	0.4404	0.4364	0.4325	0.4286	0.4247
-0.0	0.5000	0.4960	0.4920	0.4880	0.4840	0.4801	0.4761	0.4721	0.4681	0.4641

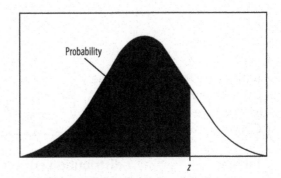

Table entry for *z* is the
probability lying below *z*.

Probability

z

Table A *(Continued)*

Z	0.00	0.01	0.02	0.03	0.04	0.05	0.06	0.07	0.08	0.09
0.0	0.5000	0.5040	0.5080	0.5120	0.5160	0.5199	0.5239	0.5279	0.5319	0.5359
0.1	0.5398	0.5438	0.5478	0.5517	0.5557	0.5596	0.5636	0.5675	0.5714	0.5753
0.2	0.5793	0.5832	0.5871	0.5910	0.5948	0.5987	0.6026	0.6064	0.6103	0.6141
0.3	0.6179	0.6217	0.6255	0.6293	0.6331	0.6368	0.6406	0.6443	0.6480	0.6517
0.4	0.6554	0.6591	0.6628	0.6664	0.6700	0.6736	0.6772	0.6808	0.6844	0.6879
0.5	0.6915	0.6950	0.6985	0.7019	0.7054	0.7088	0.7123	0.7157	0.7190	0.7224
0.6	0.7257	0.7291	0.7324	0.7357	0.7389	0.7422	0.7454	0.7486	0.7517	0.7549
0.7	0.7580	0.7611	0.7642	0.7673	0.7704	0.7734	0.7764	0.7794	0.7823	0.7852
0.8	0.7881	0.7910	0.7939	0.7967	0.7995	0.8023	0.8051	0.8078	0.8106	0.8133
0.9	0.8159	0.8186	0.8212	0.8238	0.8264	0.8289	0.8315	0.8340	0.8365	0.8389
1.0	0.8413	0.8438	0.8461	0.8485	0.8508	0.8531	0.8554	0.8577	0.8599	0.8621
1.1	0.8643	0.8665	0.8686	0.8708	0.8729	0.8749	0.8770	0.8790	0.8810	0.8830
1.2	0.8849	0.8869	0.8888	0.8907	0.8925	0.8944	0.8962	0.8980	0.8997	0.9015
1.3	0.9032	0.9049	0.9066	0.9082	0.9099	0.9115	0.9131	0.9147	0.9162	0.9177
1.4	0.9192	0.9207	0.9222	0.9236	0.9251	0.9265	0.9279	0.9292	0.9306	0.9319
1.5	0.9332	0.9345	0.9357	0.9370	0.9382	0.9394	0.9406	0.9418	0.9429	0.9441
1.6	0.9452	0.9463	0.9474	0.9484	0.9495	0.9505	0.9515	0.9525	0.9535	0.9545
1.7	0.9554	0.9564	0.9573	0.9582	0.9591	0.9599	0.9608	0.9616	0.9625	0.9633
1.8	0.9641	0.9649	0.9656	0.9664	0.9671	0.9678	0.9686	0.9693	0.9699	0.9706
1.9	0.9713	0.9719	0.9726	0.9732	0.9738	0.9744	0.9750	0.9756	0.9761	0.9767
2.0	0.9772	0.9778	0.9783	0.9788	0.9793	0.9798	0.9803	0.9808	0.9812	0.9817
2.1	0.9821	0.9826	0.9830	0.9834	0.9838	0.9842	0.9846	0.9850	0.9854	0.9857
2.2	0.9861	0.9864	0.9868	0.9871	0.9875	0.9878	0.9881	0.9884	0.9887	0.9890
2.3	0.9893	0.9896	0.9898	0.9901	0.9904	0.9906	0.9909	0.9911	0.9913	0.9916
2.4	0.9918	0.9920	0.9922	0.9925	0.9927	0.9929	0.9931	0.9932	0.9934	0.9936
2.5	0.9938	0.9940	0.9941	0.9943	0.9945	0.9946	0.9948	0.9949	0.9951	0.9952
2.6	0.9953	0.9955	0.9956	0.9957	0.9959	0.9960	0.9961	0.9962	0.9963	0.9964
2.7	0.9965	0.9966	0.9967	0.9968	0.9969	0.9970	0.9971	0.9972	0.9973	0.9974
2.8	0.9974	0.9975	0.9976	0.9977	0.9977	0.9978	0.9979	0.9979	0.9980	0.9981
2.9	0.9981	0.9982	0.9982	0.9983	0.9984	0.9984	0.9985	0.9985	0.9986	0.9986
3.0	0.9987	0.9987	0.9987	0.9988	0.9988	0.9989	0.9989	0.9989	0.9990	0.9990
3.1	0.9990	0.9991	0.9991	0.9991	0.9992	0.9992	0.9992	0.9992	0.9993	0.9993
3.2	0.9993	0.9993	0.9994	0.9994	0.9994	0.9994	0.9994	0.9995	0.9995	0.9995
3.3	0.9995	0.9995	0.9995	0.9996	0.9996	0.9996	0.9996	0.9996	0.9996	0.9997
3.4	0.9997	0.9997	0.9997	0.9997	0.9997	0.9997	0.9997	0.9997	0.9997	0.9998

Table entry for *p* and *C* is the point t^* with probability *p* lying above it and probability *C* lying between $-t^*$ and t^*.

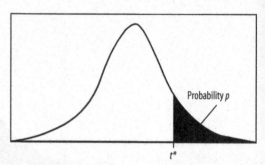

Probability *p*

t^*

Table B — *t*-distribution critical values

df	0.25	0.20	0.15	0.10	0.05	0.025	0.02	0.01	0.005	0.0025	0.001	0.0005
1	1.000	1.376	1.963	3.078	6.314	12.71	15.89	31.82	63.66	127.3	318.3	636.6
2	0.816	1.061	1.386	1.886	2.920	4.303	4.849	6.965	9.925	14.09	22.33	31.60
3	0.765	0.978	1.250	1.638	2.353	3.182	3.482	4.541	5.841	7.453	10.21	12.92
4	0.741	0.941	1.190	1.533	2.132	2.776	2.999	3.747	4.604	5.598	7.173	8.610
5	0.727	0.920	1.156	1.476	2.015	2.571	2.757	3.365	4.032	4.773	5.893	6.869
6	0.718	0.906	1.134	1.440	1.943	2.447	2.612	3.143	3.707	4.317	5.208	5.959
7	0.711	0.896	1.119	1.415	1.895	2.365	2.517	2.998	3.499	4.029	4.785	5.408
8	0.706	0.889	1.108	1.397	1.860	2.306	2.449	2.896	3.355	3.833	4.501	5.041
9	0.703	0.883	1.100	1.383	1.833	2.262	2.398	2.821	3.250	3.690	4.297	4.781
10	0.700	0.879	1.093	1.372	1.812	2.228	2.359	2.764	3.169	3.581	4.144	4.587
11	0.697	0.876	1.088	1.363	1.796	2.201	2.328	2.718	3.106	3.497	4.025	4.437
12	0.695	0.873	1.083	1.356	1.782	2.179	2.303	2.681	3.055	3.428	3.930	4.318
13	0.694	0.870	1.079	1.350	1.771	2.160	2.282	2.650	3.012	3.372	3.852	4.221
14	0.692	0.868	1.076	1.345	1.761	2.145	2.264	2.624	2.977	3.326	3.787	4.140
15	0.691	0.866	1.074	1.341	1.753	2.131	2.249	2.602	2.947	3.286	3.733	4.073
16	0.690	0.865	1.071	1.337	1.746	2.120	2.235	2.583	2.921	3.252	3.686	4.015
17	0.689	0.863	1.069	1.333	1.740	2.110	2.224	2.567	2.898	3.222	3.646	3.965
18	0.688	0.862	1.067	1.330	1.734	2.101	2.214	2.552	2.878	3.197	3.611	3.922
19	0.688	0.861	1.066	1.328	1.729	2.093	2.205	2.539	2.861	3.174	3.579	3.883
20	0.687	0.860	1.064	1.325	1.725	2.086	2.197	2.528	2.845	3.153	3.552	3.850
21	0.686	0.859	1.063	1.323	1.721	2.080	2.189	2.518	2.831	3.135	3.527	3.819
22	0.686	0.858	1.061	1.321	1.717	2.074	2.183	2.508	2.819	3.119	3.505	3.792
23	0.685	0.858	1.060	1.319	1.714	2.069	2.177	2.500	2.807	3.104	3.485	3.768
24	0.685	0.857	1.059	1.318	1.711	2.064	2.172	2.492	2.797	3.091	3.467	3.745
25	0.684	0.856	1.058	1.316	1.708	2.060	2.167	2.485	2.787	3.078	3.450	3.725
26	0.684	0.856	1.058	1.315	1.706	2.056	2.162	2.479	2.779	3.067	3.435	3.707
27	0.684	0.855	1.057	1.314	1.703	2.052	2.158	2.473	2.771	3.057	3.421	3.690
28	0.683	0.855	1.056	1.313	1.701	2.048	2.154	2.467	2.763	3.047	3.408	3.674
29	0.683	0.854	1.055	1.311	1.699	2.045	2.150	2.462	2.756	3.038	3.396	3.659
30	0.683	0.854	1.055	1.310	1.697	2.042	2.147	2.457	2.750	3.030	3.385	3.646
40	0.681	0.851	1.050	1.303	1.684	2.021	2.123	2.423	2.704	2.971	3.307	3.551
50	0.679	0.849	1.047	1.299	1.676	2.009	2.109	2.403	2.678	2.937	3.261	3.496
60	0.679	0.848	1.045	1.296	1.671	2.000	2.099	2.390	2.660	2.915	3.232	3.460
80	0.678	0.846	1.043	1.292	1.664	1.990	2.088	2.374	2.639	2.887	3.195	3.416
100	0.677	0.845	1.042	1.290	1.660	1.984	2.081	2.364	2.626	2.871	3.174	3.390
1000	0.675	0.842	1.037	1.282	1.646	1.962	2.056	2.330	2.581	2.813	3.098	3.300
	0.674	0.841	1.036	1.282	1.645	1.960	2.054	2.326	2.576	2.807	3.091	3.291
	50%	60%	70%	80%	90%	95%	96%	98%	99%	99.5%	99.8%	99.9%

Confidence level *C*

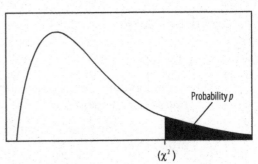

Table entry for *p* is the point (χ^2) with probability *p* lying above it.

(χ^2)

Probability *p*

Table C $\qquad$ χ^2-critical values

df	Tail probability *p*										
	0.25	0.20	0.15	0.10	0.05	0.025	0.02	0.01	0.005	0.0025	0.001
1	1.32	1.64	2.07	2.71	3.84	5.02	5.41	6.63	7.88	9.14	10.83
2	2.77	3.22	3.79	4.61	5.99	7.38	7.82	9.21	10.60	11.98	13.82
3	4.11	4.64	5.32	6.25	7.81	9.35	9.84	11.34	12.84	14.32	16.27
4	5.39	5.99	6.74	7.78	9.49	11.14	11.67	13.28	14.86	16.42	18.47
5	6.63	7.29	8.12	9.24	11.07	12.83	13.39	15.09	16.75	18.39	20.51
6	7.84	8.56	9.45	10.64	12.59	14.45	15.03	16.81	18.55	20.25	22.46
7	9.04	9.80	10.75	12.02	14.07	16.01	16.62	18.48	20.28	22.04	24.32
8	10.22	11.03	12.03	13.36	15.51	17.53	18.17	20.09	21.95	23.77	26.12
9	11.39	12.24	13.29	14.68	16.92	19.02	19.68	21.67	23.59	25.46	27.88
10	12.55	13.44	14.53	15.99	18.31	20.48	21.16	23.21	25.19	27.11	29.59
11	13.70	14.63	15.77	17.28	19.68	21.92	22.62	24.72	26.76	28.73	31.26
12	14.85	15.81	16.99	18.55	21.03	23.34	24.05	26.22	28.30	30.32	32.91
13	15.98	16.98	18.20	19.81	22.36	24.74	25.47	27.69	29.82	31.88	34.53
14	17.12	18.15	19.41	21.06	23.68	26.12	26.87	29.14	31.32	33.43	36.12
15	18.25	19.31	20.60	22.31	25.00	27.49	28.26	30.58	32.80	34.95	37.70
16	19.37	20.47	21.79	23.54	26.30	28.85	29.63	32.00	34.27	36.46	39.25
17	20.49	21.61	22.98	24.77	27.59	30.19	31.00	33.41	35.72	37.95	40.79
18	21.60	22.76	24.16	25.99	28.87	31.53	32.35	34.81	37.16	39.42	42.31
19	22.72	23.90	25.33	27.20	30.14	32.85	33.69	36.19	38.58	40.88	43.82
20	23.83	25.04	26.50	28.41	31.41	34.17	35.02	37.57	40.00	42.34	45.31
21	24.93	26.17	27.66	29.62	32.67	35.48	36.34	38.93	41.40	43.78	46.80
22	26.04	27.30	28.82	30.81	33.92	36.78	37.66	40.29	42.80	45.20	48.27
23	27.14	28.43	29.98	32.01	35.17	38.08	38.97	41.64	44.18	46.62	49.73
24	28.24	29.55	31.13	33.20	36.42	39.36	40.27	42.98	45.56	48.03	51.18
25	29.34	30.68	32.28	34.38	37.65	40.65	41.57	44.31	46.93	49.44	52.62
26	30.43	31.79	33.43	35.56	38.89	41.92	42.86	45.64	48.29	50.83	54.05
27	31.53	32.91	34.57	36.74	40.11	43.19	44.14	46.96	49.64	52.22	55.48
28	32.62	34.03	35.71	37.92	41.34	44.46	45.42	48.28	50.99	53.59	56.89
29	33.71	35.14	36.85	39.09	42.56	45.72	46.69	49.59	52.34	54.97	58.30
30	34.80	36.25	37.99	40.26	43.77	46.98	47.96	50.89	53.67	56.33	59.70
40	45.62	47.27	49.24	51.81	55.76	59.34	60.44	63.69	66.77	69.70	73.40
50	56.33	58.16	60.35	63.17	67.50	71.42	72.61	76.15	79.49	82.66	86.66
60	66.98	68.97	71.34	74.40	79.08	83.30	84.58	88.38	91.95	95.34	99.61
80	88.13	90.41	93.11	96.58	101.9	106.6	108.1	112.3	116.3	120.1	124.8
100	109.1	111.7	114.7	118.5	124.3	129.6	131.1	135.8	140.2	144.3	149.4

IV. Symbol Usage

A. Capitalization

1. Capital letters refer to population parameters. p refers to a population proportion.

2. Lowercase letters refer to statistics. $\hat{p}$ refers to a sample proportion.

3. X refers to a set of population elements; and x, to a set of sample elements, while N refers to population size; and n, to sample size.

B. Greek versus Roman Letters

1. Greek letters refer to population attributes. μ refers to a population mean and σ refers to the standard deviation of a population

2. Roman letters refer to samples. $\bar{x}$ refers to a sample mean, and σ to the standard deviation of a sample.

C. Commonly used symbols

1. $\bar{x}$ refers to a sample mean.

2. s refers to the standard deviation of a sample.

3. s^2 refers to the variance of a sample.

4. r is the sample correlation coefficient.

D. Simple Linear Regression

1. b_0 is the intercept constant in a population regression line.

2. b_1 is the regression coefficient (slope) in a population regression line.

3. r^2 refers to the coefficient of determination.

4. b_0 is the intercept constant in a sample regression line.

5. b_1 refers to the regression coefficient in a sample regression line (i.e., the slope).

6. s_{b1} refers to the standard error of the slope of a regression line.

E. Probability

1. $P(A)$ defines the probability that event A will occur.

2. $P(A \mid B)$ defines the conditional probability that event A occurs, given that event B has taken place.

3. $P(A^c)$ defines the probability of the complement of event A.

4. $P(A \cap B)$ defines the probability of the intersection of events A and B.

5. $P(A \cup B)$ defines the probability of the union of events A and B.

6. $A \cap B$ defines the intersection of events A and B.

7. $A \cup B$ defines the union of events A and B.

8. $\{A,B,C\}$ defines the set of elements consisting of A, B, and C.

9. $\{\varnothing\}$ refers to the null set.

F. Hypothesis Testing/Statistical Inference

1. H_0 defines null hypothesis.

2. H_a refers to an alternative hypothesis.

3. α refers to the significance level.

4. β refers to the probability of committing a Type II error.

5. z refers to a standardized score, also known as a z-score.

6. X^2 refers to a chi-square statistic.

G. Statistical Math Symbols

1. Σ defines the summation symbol.

2. Σx or Σx_i defines the sum of a set of n observations. For example,
$$\Sigma x_i = \Sigma x = x_1 + x_2 + \ldots + x_n.$$

3. Var(X) defines the variance of the random variable X.

4. SE defines the standard error of a statistic.

5. ME defines the margin of error.

6. DF defines the degrees of freedom $(n - 1)$.

Glossary

Addition rule: $P(A \cup B) = P(A) + P(A) - P(A \cap B)$ aids in computing the chances of one of several events occurring at a given time.

Alpha (α): the probability of a Type I error. See significance level.

Alternative hypothesis: the hypothesis stating what the researcher is seeking evidence of. A statement of inequality. It can be written looking for the difference or change in one direction from the null hypothesis or both.

Association: relationship between or among variables.

Back-transform: the process by which values are substituted into a model of transformed data, and then reversing the transforming process to obtain the predicted value or model for nontransformed data.

Bar chart: a graphical display used with categorical data, where frequencies for each category are shown in vertical bars.

Bell-shaped: often used to describe the normal distribution. See **mound-shaped**.

Beta (β): The probability of a Type II error. See **power**.

Bias: the term for systematic deviation from the truth (parameter), caused by systematically favoring some outcomes over others.

Biased: a sampling method is biased if it tends to produce samples that do not represent the population.

Bimodal: a distribution with two clear peaks.

Binomial distribution: the probability distribution of a binomial random variable.

Binomial random variable: a random variable X (a) that has a fixed number of trials of a random phenomenon n, (b) that has only two possible outcomes on each trial, (c) for which the probability of a success is constant for each trial, and (d) for which each trial is independent of other trials.

Bins: the intervals that define the "bars" of a histogram.

Bivariate data: consists of two variables, an explanatory and a response variable, usually quantitative.

Blinding: practice of denying knowledge to subjects about which treatment is imposed upon them.

Blocks: subgroups of the experimental units that are separated by some characteristic before treatments are assigned because they may respond differently to the treatments.

Box-and-whisker plot/boxplot: a graphical display of the five-number summary of a set of data, which also shows outliers.

Categorical variable: a variable recorded as labels, names, or other non-numerical outcomes.

Census: a study that observes, or attempts to observe, every individual in a population.

Central limit theorem: as the size n of a simple random sample increases, the shape of the sampling distribution of $\bar{x}$ tends toward being normally distributed.

Chance device: a mechanism used to determine random outcomes.

Cluster sample: a sample in which a simple random sample of heterogeneous subgroups of a population is selected.

Clusters: heterogeneous subgroups of a population.

Coefficient of determination (r^2): percent of variation in the response variable explained by its linear relationship with the explanatory variable.

Complement: the complement of an event is that event not occurring.

Complementary events: two events whose probabilities add up to 1.

Completely randomized design: one in which all experimental units are assigned treatments solely by chance.

Conditional distribution: see conditional frequencies.

Conditional frequencies: relative frequencies for each cell in a two-way table relative to one variable.

Conditional probability: the probability of an event occurring given that another has occurred. The probability of A given that B has occurred is denoted as $P(A|B)$.

Confidence intervals: give an estimated range that is likely to contain an unknown population parameter.

Confidence level: the level of certainty that a population parameter exists in the calculated confidence interval.

Confounding: the situation where the effects of two or more explanatory variables on the response variable cannot be separated.

Confounding variable: a variable whose effect on the response variable cannot be untangled from the effects of the treatment.

Contingency table: see **two-way table**.

Continuous random variables: those typically found by measuring, such as heights or temperatures.

Control group: a baseline group that may be given no treatment, a faux treatment like a placebo, or an accepted treatment that is to be compared to another.

Control: the principle that potential sources of variation due to variables not under consideration must be reduced.

Convenience sample: composed of individuals who are easily accessed or contacted.

Correlation coefficient (r): a measure of the strength of a linear relationship,

$$r = \frac{1}{n-1} \sum \left(\frac{x_i - \bar{x}}{s_x} \right) \left(\frac{y_i - x\bar{y}}{s_y} \right).$$

Critical value: the value that the test statistic must exceed in order to reject the null hypothesis. When computing a confidence interval, the value of t^* (or z^*) where $\pm t^*$ (or $\pm z^*$) bounds the central $C\%$ of the t (or z) distribution.

Cumulative frequency: the sums of the frequencies of the data values from smallest to largest.

Data set: collection of observations from a sample or population.

Dependent events: two events are called dependent when they are related and the fact that one event has occurred changes the probability that the second event occurs.

Discrete random variables: those usually obtained by counting.

Disjoint events: events that cannot occur simultaneously.

Distribution: frequencies of values in a data set.

Dotplot: a graphical display used with univariate data. Each data point is shown as a dot located above its numerical value on the horizontal axis.

Double-blind: when both the subjects and data gatherers are ignorant about which treatment a subject received.

Empirical rule (68-95-99.7 rule): gives benchmarks for understanding how probability is distributed under a normal curve. In the normal distribution, 68% of the observations are within one standard deviation of the mean, 95% is within two standard deviations of the mean, and 99.7% is within three standard deviations of the mean.

Estimation: the process of determining the value of a population parameter from a sample statistic.

Expected value: the mean of a probability distribution.

Experiment: a study where the researcher deliberately influences individuals by imposing conditions and determining the individuals' responses to those conditions.

Experimental units: individuals (a person, a plot of land, a machine, or any single material unit) in an experiment.

Explanatory variable: explains the response variable, sometimes known as the treatment variable.

Exponential model: a model of the form of $y = ab^x$.

Extrapolation: using a model to predict values far outside the range of the explanatory variable, which is prone to creating unreasonable predictions.

Factors: one or more explanatory variables in an experiment.

First quartile: symbolized Q_1, represents the median of the lower 50% of a data set.

Five-number summary: the minimum, first quartile (Q_1), median, third quartile (Q_3), and maximum values in a data set.

Frequency table: a display organizing categorical or numerical data and how often each occurs.

Geometric distribution: the probability distribution of a geometric random variable X. All possible outcomes of X before the first success is seen and their associated probabilities.

Geometric random variable: a random variable X that (a) has two possible outcomes in each trial, (b) for which the probability of a success is constant for each trial, and (c) for which each trial is independent of the other trials.

Graphical display: a visual representation of a distribution.

Histogram: used with univariate data, frequencies are shown on the vertical axis, and intervals or bins define the values on the horizontal axis.

Independent events: two events are called independent when knowing that one event has occurred does not change the probability that the second event occurs.

Independent random variables: if the values of one random variable have no association with the values of another, the two variables are called independent random variables.

Influential point: an extreme value whose removal would drastically change the slope of the least-squares regression model.

Interquartile range: describes the spread of the middle 50% of a data set, $IQR = Q_3 - Q_1$.

Joint distribution: see **joint frequencies**.

Joint frequencies: frequencies for each cell in a two-way table relative to the total number of data.

Law of large numbers: the long-term relative frequency of an event gets closer to the true relative frequency as the number of trials of a random phenomenon increases.

Least-squares regression line (LSRL): the "best-fit" line that is calculated by minimizing the sum of the squares of the differences between the observed and predicted values of the line. The LSRL has the equation $\hat{y} = b_0 + b_1 x$.

Levels: the different quantities or categories of a factor in an experiment.

Linear regression: a method of finding the best model for a linear relationship between the explanatory and response variable.

Logarithmic transformation: procedure that changes a variable by taking the logarithm of each of its values.

Lurking variable: a variable that has an effect on the outcome of a study but was not part of the investigation.

Margin of error: a range of values to the left and right of a point estimate.

Marginal distribution: see **marginal frequencies**.

Marginal frequencies: row totals and column totals in a two-way table.

Matched-pairs design: the design of a study where experimental units are naturally paired by a common characteristic, or with themselves in a before–after type of study.

Maximum: the largest numerical value in a data set.

Mean: the arithmetic average of a data set; the sum of all the values divided by the number of values, $\bar{x} = \dfrac{\sum x_i}{n}$.

Mean of a binomial random variable X: $\mu x = np$.

Mean of a discrete random variable: $\mu_x = \sum\limits_{i=1}^{n} x_i P(x_i)$.

Mean of a geometric random variable: $\mu_x = \dfrac{1}{p}$.

Measures of center: these locate the middle of a distribution. The mean and median are measures of center.

Median: the middle value of a data set; the equal areas point, where 50% of the data are at or below this value, and 50% of the data are at or above this value.

Minimum: the smallest numerical value in a data set.

Mound-shaped: resembles a hill or mound; a distribution that is symmetric and unimodal.

Multiplication rule: $P(A \cap B) = P(A) \cdot P(B \mid A)$ is used when we are interested in the probability of two events occurring simultaneously, or in succession.

Multistage sample: a sample resulting from multiple applications of cluster, stratified, and/or simple random sampling.

Mutually exclusive events: see **disjoint events**.

Nonresponse bias: the situation where an individual selected to be in the sample is unwilling, or unable, to provide data.

Normal distribution: a continuous probability distribution that appears in many situations, both natural and man-made. It has a bell-shape and the area under the normal density curve is always equal to 1.

Null hypothesis: the hypothesis of no difference, no change, and no association. A statement of equality, usually written in the form H_0: *parameter* = hypothesized value.

Observational study: attempts to determine relationships between variables, but the researcher imposes no conditions as in an experiment.

Observed values: actual outcomes or data from a study or an experiment.

One-way table: a frequency table of one variable.

Outlier: an extreme value in a data set. Quantified by being less than $Q_1 - 1.5IQR$ or more than $Q_3 + 1.5IQR$.

Percentiles: divide the data set into 100 equal parts. An observation at the Pth percentile is higher than P percent of all observations.

Placebo: a faux treatment given in an experiment that resembles the real treatment under consideration.

Placebo effect: a phenomenon where subjects show a response to a treatment merely because the treatment is imposed regardless of its actual effect.

Point estimate: an approximate value that has been calculated for the unknown parameter.

Population: the collection of all individuals under consideration in a study.

Population parameter: a characteristic or measure of a population.

Position: location of a data value relative to the population.

Power: the probability of correctly rejecting the null hypothesis when it is in fact false. Equal to $1 - \beta$. See beta and Type II error.

Power model: a function in the form of $y = ax^b$.

Predicted value: the value of the response variable predicted by a model for a given explanatory variable.

Probability: describes the chance that a certain outcome of a random phenomenon will occur.

Probability distribution: a discrete random variable X is a function of all n possible outcomes of the random variable (x_i) and their associated probabilities $P(x_i)$.

Probability sample: composed of individuals selected by chance.

P-value: the probability of observing a test statistic as extreme as, or more extreme than, the statistic obtained from a sample, under the assumption that the null hypothesis is true.

Quantitative: a variable whose values are counts or measurements.

Random digit table: a chance device that is used to select experimental units or conduct simulations.

Random phenomena: those outcomes that are unpredictable in the short term, but nevertheless, have a long-term pattern.

Random sample: a sample composed of individuals selected by chance.

Random variables: numerical outcome of a random phenomenon.

Randomization: the process by which treatments are assigned by a chance mechanism to the experimental units.

Randomized block design: first, units are sorted into subgroups or blocks, and then treatments are randomly assigned within the blocks.

Range: calculated as the maximum value minus the minimum value in a data set.

Relative frequency: percentage or proportion of the whole number of data.

Relative frequency segmented bar chart: a method of graphing a conditional distribution.

Replication: the practice of reducing chance variation by assigning each treatment to many experimental units.

Residual: observed value minus predicted value of the response variable.

Response bias: because of the manner in which an interview is conducted, due to the phrasing of questions, or because of the attitude of the respondent, inaccurate data are collected.

Response variable: measures the outcomes that have been observed.

Sample: a selected subset of a population from which data are gathered.

Sample statistic: result of a sample used to estimate a parameter.

Sample survey: a study that collects information from a sample of a population in order to determine one or more characteristics of the population.

Sampling distribution: the probability distribution of a sample statistic when a sample is drawn from a population.

Sampling distribution of the sample mean $\bar{x}$: the distribution of sample means from all possible simple random samples of size n taken from a population.

Sampling distribution of a sample proportion $\hat{p}$: the distribution of sample proportions from all possible simple random samples of size n taken from a population.

Sampling error: see **sampling variability**.

Sampling variability: natural variability due to the sampling process. Each possible random sample from a population will generate a different sample statistic.

Scatterplots: used to visualize bivariate data. The explanatory variable is shown on the horizontal axis and the response variable is shown on the vertical axis.

Significance level: the probability of a Type I error. A benchmark against which the *P*-value compared to determine if the null hypothesis will be rejected. See also **alpha (α)**.

Simple random sample (SRS): a sample where n individuals are selected from a population in a way that every possible combination of n individuals is equally likely.

Simulation: a method of modeling chance behavior that accurately mimics the situation being considered.

Skewed: a unimodal, asymmetric, distribution that tends to slant—most of the data are clustered on one side of the distribution and "tails" off on the other side.

Standard deviation of a binomial random variable X: $\sigma_x = \sqrt{np(1-p)}$.

Standard deviation of a discrete random variable X: $\sigma_x = \sqrt{\sigma_x^2}$.

Standard deviation: used to measure variability of a data set. It is calculated as the square root of the variance of a set of data,

$$s = \sqrt{\frac{\Sigma(x_i - \bar{x})^2}{n-1}}.$$

Standard error: an estimate of the standard deviation of the sampling distribution of a statistic.

Standard normal probabilities: the probabilities calculated from values of the standard normal distribution.

Standardized score: the number of standard deviations an observation lies from the mean, $z = \dfrac{\text{observation} - \text{mean}}{\text{standard deviation}}$.

Statistically significant: when a sample statistic is shown to be far from a hypothesized parameter. When the P-value is less than the significance level.

Stemplot: also called a stem-and-leaf plot. Data are separated into a stem and a leaf by place value and organized in the form of a histogram.

Strata: subgroups of a population that are similar or homogenous.

Stratification: part of the sampling process where units of the study are separated into strata.

Stratified random sample: a sample in which simple random samples are selected from each of several homogenous subgroups of the population, known as strata.

Subjects: individuals in an experiment that are people.

Symmetric: the distribution that resembles a mirror image on either side of the center.

Systematic random sample: a sample where every kth individual is selected from a list or queue.

Test statistic: the number of standard deviations (standard errors) that a sample statistic lies from a hypothesized population parameter.

Third quartile: symbolized Q_3, represents the median of the upper 50% of a data set.

Transformation: changing the values of a data set using a mathematical operation.

Treatments: combinations of different levels of the factors in an experiment.

Two-way table: a frequency table that displays two categorical variables.

Type I error: rejecting a null hypothesis when it is in fact true.

Type II error: failing to reject a null hypothesis when it is in fact false.

Undercoverage: when some individuals of a population are not included in the sampling process.

Uniform: all data values in the distribution have similar frequencies.

Unimodal: a distribution with a single, clearly defined, peak.

Univariate: one-variable data.

Variables: characteristics of the individuals under study.

Variability: the spread in a data set.

Variance: used to measure variability, the average of the squared deviations from the mean, $s_x^2 = \sqrt{\dfrac{\Sigma(x_i - \bar{x})^2}{n-1}}$.

Variance of a binomial random variable X: $\sigma_x^2 = np(1-p)$.

Variance of a discrete random variable X: $\sigma_x^2 = \sum_{i=1}^{n}(x_i - \mu_x)^2 \circ P(x_i)$.

Venn diagram: graphical representation of sets or outcomes and how they intersect.

Voluntary response bias: bias due to the manner in which people choose to respond to voluntary surveys.

Voluntary response sample: composed of individuals who choose to respond to a survey because of interest in the subject.

z-score: see **standardized score**.

Notes

Notes

Notes

Notes

Notes

Notes

Notes

Notes

Notes